ELEMENTARY FOREST SURVEYING

AND MAPPING

II

R. L. Wilson

Published by
O.S.U. Book Stores, Inc.
Corvallis, Oregon
1978
Litho--U.S.A.

Revised Edition
1982

ISBN No. 0-88246-136-2

Preface

This text was written for the purpose of explaining the use of the engineer's transit, theodolite, engineer's level and plane table in the solution of fundamental problems related to Forest Engineering. While the use of the infrared and electronic distance measuring instruments do have a place in Forest Engineering, they have been omitted as has been celestial astronomy, photogrammetrey and public land survey. It is necessary to master the fundamentals before attempting to solve more advanced problems with more sophisticated equipment.

Logarithmic tables have also been omitted from the text because it is believed that they have outlived their usefulness in the age of the electric calculators.

Table of Contents

Figure   Page

ELEMENTARY FOREST SURVEYING
AND MAPPING

INTRODUCTION

Surveying is an art and an art is defined as the power of performing certain activities as acquired by experience, study or observation. Perhaps the last word "observation" should be omitted in the description of surveying because it is best learned by doing and studying.

Surveying is the application of the skill of measuring and locating lines, angles and elevations on or near the surface of the earth or determining the relative position of points to each other.

Surveying is divided into field work, which consists of obtaining data, and office work, which is the computation and drafting or converting field measurements to serve the purpose of the survey. Most surveying is done for one of two reasons: (1) to establish or re-establish the boundaries of land and (2) to provide information necessary for some type of construction. The first is referred to as land surveying and the second is called construction surveying.

When the survey is extended over a large portion of the earth's surface, whose shape is a spheroid of revolution, the curvature of the earth must be taken into consideration. This is called geodetic surveying. When the area being surveyed is relatively small and the area is considered to be a plane surface with no curvature involved, it is referred to as plane surveying. It is the latter system with which this text is concerned.

Surveys must be correct. There is no acceptance or tolerance of human error in the field or in the office. Employers do not look favorably on mistakes which result in requiring twice the time to complete the job correctly or which result in more costly changes at a later date. It is up to the surveyor in the field, therefore, to keep his field notes to the best of his ability. The notes are just as important as the data itself.

The method in which the notes are taken is really a matter of personal preference and the author prefers to start all notes at the bottom of the note page and go up for the reason that in traversing, the data and sketch will agree line by line. It is also very easy to visualize just where you are on the traverse by holding the note book in front of you and sighting ahead along the line to be followed. However, all field notes for any kind of leveling or stadia are started at the top of the page. Irrespective of the form in which the notes are taken, the surveyor must do the following:

(1) Record all field notes carefully in the field book. Never erase but draw a line through the incorrect value and write the correct value above or below and circle the correct value.

(2) Check all data before recording.

(3) Record all data legibly in an orderly manner so that it may be used for office computation by others if necessary.

(4) Use sketches to clarify all data.

The meaning of two words, accuracy and precision should be made clear. Precision means closeness in agreement between several measurements and accuracy means correctness or freedom from mistakes and carelessness. Precision is meaningless unless accuracy is maintained. A line measured in two directions for example is found to be 345.54 and 344.55 feet. The hundredths of feet would indicate precision in both measurements. However, the difference of 1 foot between the two measurements indicates that carelessness has not been eliminated and therefore one of the measurements was incorrect and time was wasted in obtaining precision that is useless.

No measurement is perfect but the accuracy and precision will improve as more care is taken in obtaining the data. Cost also goes up as more time is required to insure precision and to increase the accuracy. The surveyor must know the purpose and the intended use of the survey in order to select the proper equipment for the job and obtain the needed field data with compatible costs. The table below will give some idea regarding the accuracy associated with various types of surveying instruments.

Crew Size	Direction	Distance	Vertical Difference in Elevation	Accuracy Relative Error
1	Hand Compass	Pacing	Aneroid Barometer	1:80
2 or 3	Staff Compass	Topograpnic Tape	Clinometer of Abey	1:800
3	Transit	Engineer's Tape	Vertical Angles Measured to 1'	1:3500
3	Theodolite	Engineer's Tape	Vertical Angles Measured to 6"	1:10,000
4	Theodolite	E.D.M.		1:50,000

TAPES

Tapes made for determining distances come in a variety of lengths, weights and materials. The words tape and chain should not be confused. The term tape is used to indicate the instrument with which distances are determined. The word chain indicates that unit distance of 66 feet or 100 links and chaining is the act of measuring either horizontal or slope distance between two points. The surveyor's tape has the same units of measurement as did the Gunter chain which was a series of wire links fastened together (Figure 1-1) The surveyor's tape is usually used with the staff compass and abney and is graduated in chains and links. It also has two trailers which enable the engineer to set points which are one or two chains horizontal distance apart while measuring slope distance. The one chain trailer calibrations are placed on the underside of the tape (Figure 1-2). As the metric system is adopted, more tapes will be used which are calibrated in meters, decimeters and centimeters with the end decimeter graduated in millimeters.

Tapes used with transits and theodolites are called engineer's tapes and come in length of 100, 200 and 300 feet. Most tapes are made of steel and the weight and width vary with the length, the longer tapes being narrower and lighter. There are some fiberglass tapes used which are very durable but stretch when tension is applied. The engineer's tapes are made in two different styles. One tape is known as an adding tape which may be 100 or 200 feet in length and one foot of the tape in addition to the 100 or 200 foot length is graduated into 10ths and sometimes 100ths of feet. The other tape is known as a subtracting tape and it also may be 100 or 200 feet long. The first foot of these tapes between the zero mark and the one foot mark is sub-divided into 10ths or 100ths of feet. The name is derived from the fact that the distance measured is determined by subtracting the fractional part of a foot held by the head chainman from the mark that the rear chainman is holding. Figure 1-3 illustrates the graduations on an adding and subtracting tape. Some tapes which are 150 feet in length are known as Invar tapes and are made of a composition of nickel and steel and have a very low coefficient of thermal expansion and are used only where very precise measurements are needed.

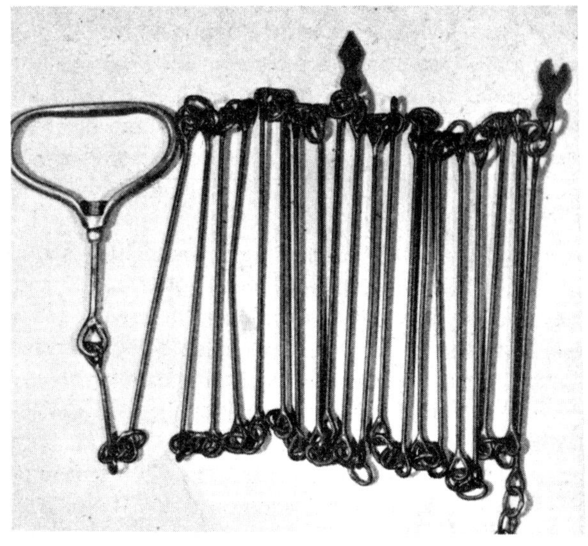

Figure 1-1. Gunter chain.

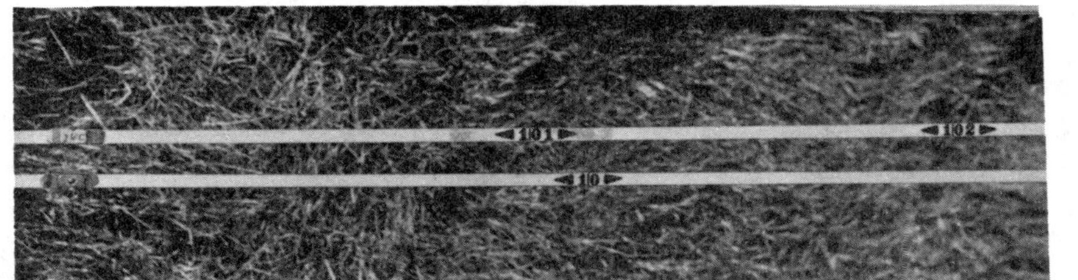

Figure 1-2. Upper and lower side of topographic tape showing graduations and one chain trailer.

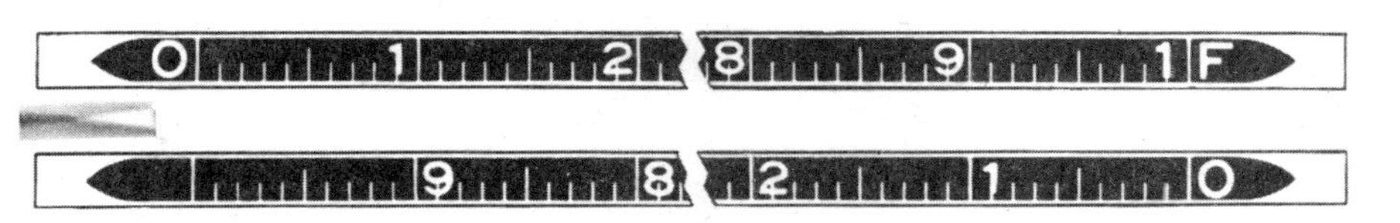

Figure 1-3. Graduation of end foot on an Engineer's adding and subtracting tape.

In forest surveying, the tape is very seldom rolled up on a reel but is coiled into a figure eight in the following manner. Figure 1-4 shows the topographic tape being coiled beginning at the zero end of the tape. The engineer's tape may be coiled beginning at either end of the tape (Figure 1-5). Facing the zero end and with your back to the other end of the tape which should lie in approximately a straight line behind you, place the zero end or the 100 foot mark of the tape in the palm of the left hand which is facing up (Figure 1-6). With the right hand holding the tape loosely, the fingers pointed toward the body, extend both arms, left arm forward and right arm back along the line of the tape, allowing the tape to slide freely through the fingers of the right hand until the right hand reaches the 5 foot mark on the tape.

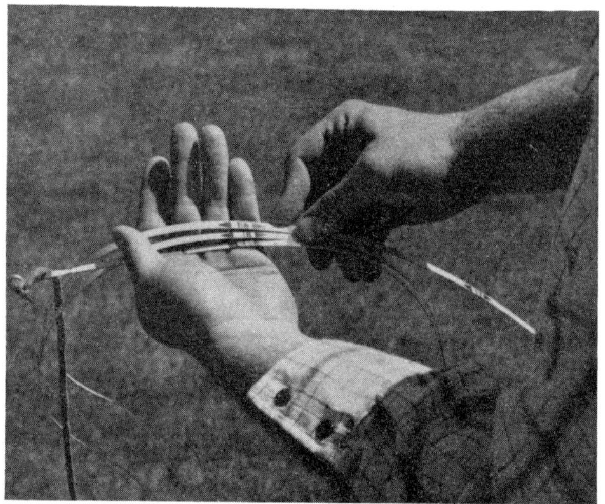

Figure 1-4. Coiling a topographic tape.

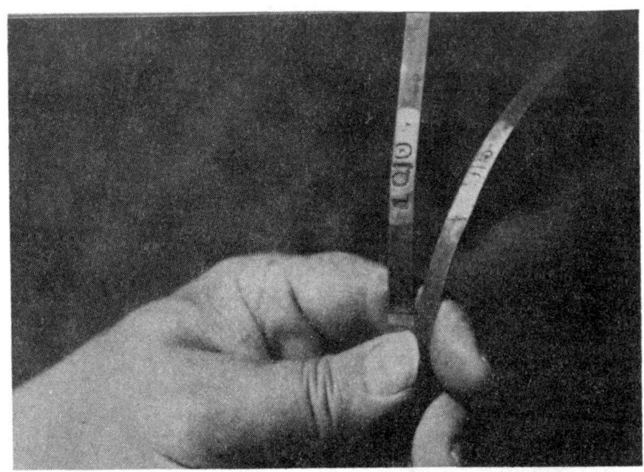

Figure 1-5. Five foot loop of engineer's tape.

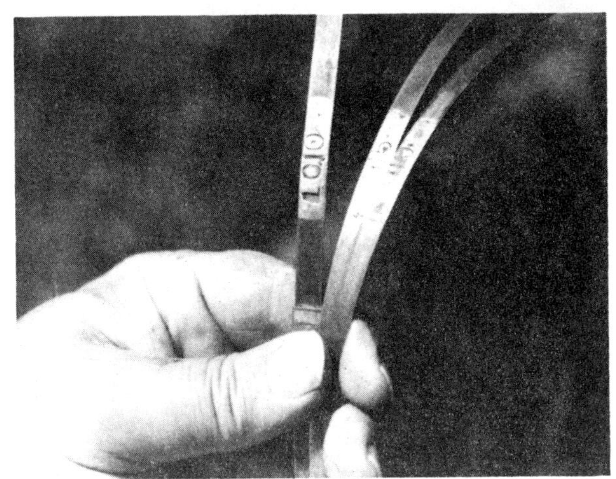

Figure 1-6. Coils of an engineer's tape.

The tape is then grasped by the right hand and both hands are then brought together so that the 5 foot mark is placed on the zero mark (95 foot mark on the 100 foot mark) of the tape, being careful not to let the tape turn or twist in the right hand. The topographic tape is coiled in loops of 8 links (Figure 1-7). This process is repeated for the next 5 feet of tape and thus the 10 foot mark is placed on the 5 foot mark, etc., until the entire length of the tape has been coiled into 5 foot loops or 8 link loops which have the appearance of figure eights (Figure 1-8).

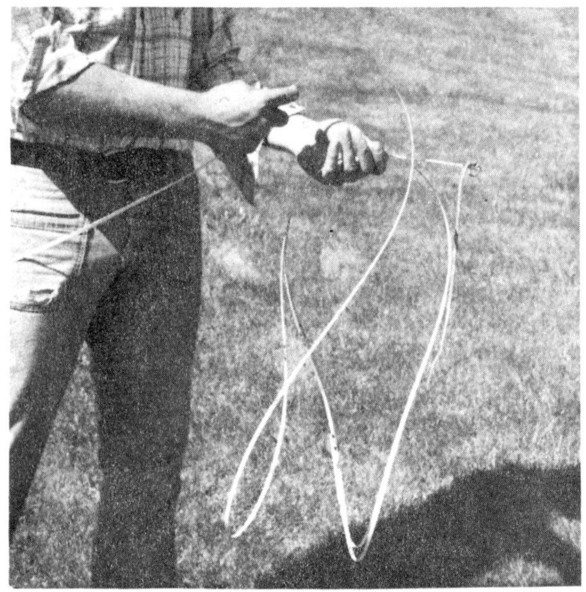

Figure 1-7. Coiling a topographic tape.

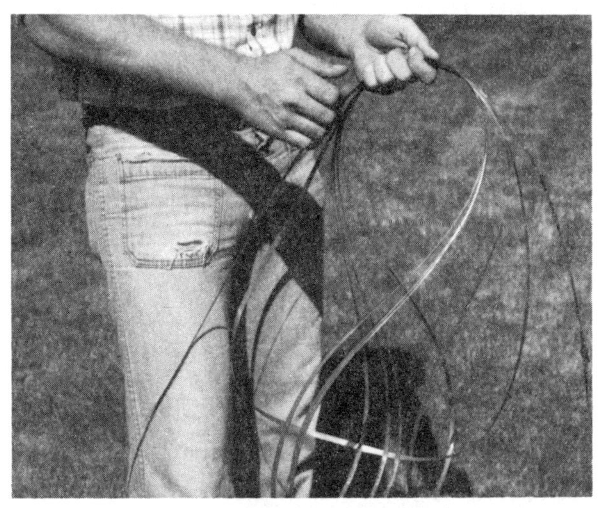

Figure 1-8. Figure eights of a coiled tape.

The leather thongs at the end of the tape are now wrapped about the tape being sure to wrap the thongs back onto the ends of the tape. The thongs are made secure by pulling the end of the thong through the layers of the coiled tape. The tape, which now will assume a form similar to a figure eight, is held in a vertical position in front of the body (Figure 1-9). Grasping that part of the figure eight which is closest to the body with the right hand, and the part which is farthest away with the left hand, the tape is now twisted into a circular form by turning the wrists in opposite directions (Figures 1-10, 1-11, 1-12, 1-13).

To lay the tape out, the reverse procedure is followed. The circle loop is twisted to return the tape to the figure

Figure 1-10. Throwing figure eight into small coil.

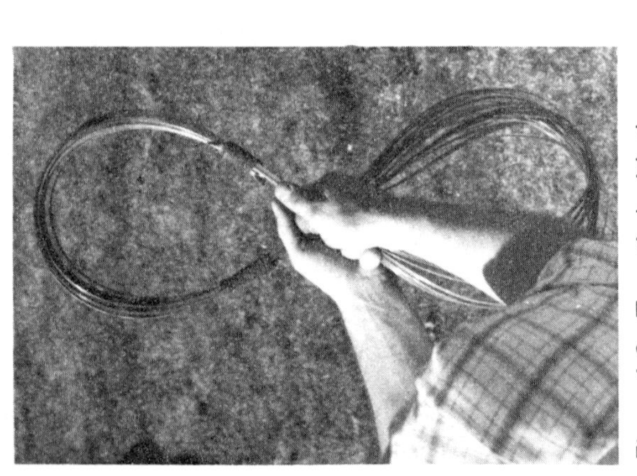

Figure 1-9. Tape tied with thongs.

Figure 1-11. Throwing figure eight into small coil.

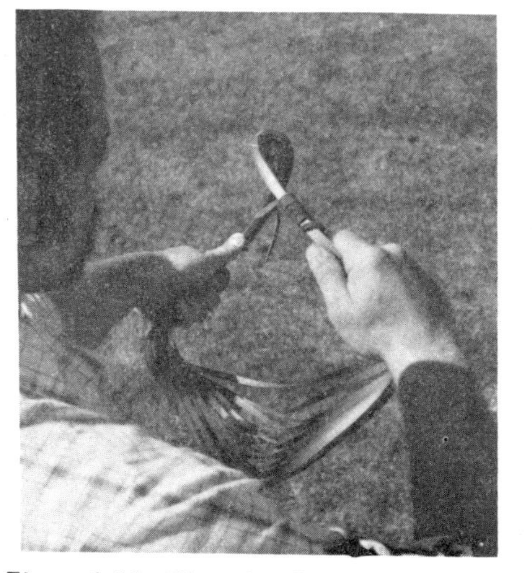

Figure 1-12. Throwing figure eight into small coil.

Figure 1-13. Throwing figure eight into small coil.

eight form. The thongs are untied and holding the tape
the left hand, being sure to keep the coiled loops in th
same order they were taken in, lay the tape out on th
ground beginning with the top coil. Make sure that ea
coil is taken off away from the body rather than toward
The latter way will result in the tape being twisted into
spiral which could cause the tape to break when tension
applied. Also make certain when uncoiling the tape th
the zero end is the last part laid out because the head cha
man always takes the zero end ahead with him. In order
do this, it may be necessary to turn the tape inside c
before loosening the thongs and uncoiling the tape. It is n
always easy to turn the tape end for end in heavy brush
enable the head chainman to take the zero end of the ta

A second way of throwing a tape is the basket weave
ere the tape is coiled in the shape of a figure eight by
ucing a half twist in the tape for every loop coiled. It
y be coiled by looping one loop around your foot. The
ps may be of any length so long as they are all uniform.

A third method is to coil the tape into flat loops that are
feet in length. This is done by giving each loop a full
st of 360°. This method will result in the tape being
ned into a perfect circle and this circle is then coiled
 a small circular coil by making 3 small loops. To take
 tape down, two people are needed. Since a full twist
 given to each loop, it is taken down by rolling the
re loop, not unlike line coming off a level wind fishing
.

The process of chaining is the same whether slope dis-
e or horizontal distance is being measured. If the mark
e held is either the 100 foot or 200 foot mark, the
ng may be wrapped about one hand in such a manner
 the fingers can grasp the tape. The rear chainman faces
 tape with his feet spread well apart, holding the tape
e to his body about waist high but in a manner which
 not interfere with the plumb bob. In this manner, ten-
 on the tape is created with the body and legs rather
 with the arms only. The plumb bob string is placed
 the tape and clamped in place on the mark with the
nb and index finger on the other hand. If the mark
g held by the rear chainman is elsewhere on the tape,
plumb bob string is again clamped on a foot mark only
 the tape is held by the other hand after being pulled
nd behind the back (Figures 1-14, 1-15). This enables
chainman to provide tension on the tape with his legs

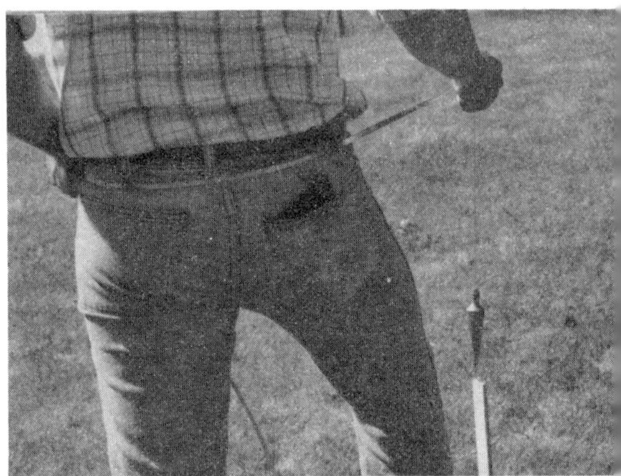

Figures 1-14, 1-15. Rear chainman holding tape at p
other than at the end of the t

and prevents creating a small radius on the tape as would happen if the tension were created with the hands or arms only. If one end of the tape needs to be held higher in order to flat chain, the tape may be placed over one shoulder (Figures 1-16, 1-17). Proper tension is difficult to create and maintain if the arms have to be used above shoulder height.

Figure 1-16. Rear chainman chaining with tape at shoulder height.

Figure 1-17. Rear chainman chaining with tape at shoulder
height.

The head chainman follows the same procedure: faces
the tape, wraps the thong around one hand while the other
hand is free to move the plumb bob string along the tape
until the plumb bob is over the hub tack or point being
measured to. He can then clamp the string onto the tape

and observe the 10ths or 100ths of feet without having to plumb the point and read the distance at the same time (Figures 1-18, 1-19).

Figure 1-18. Head chainman measuring at zero end of tape.

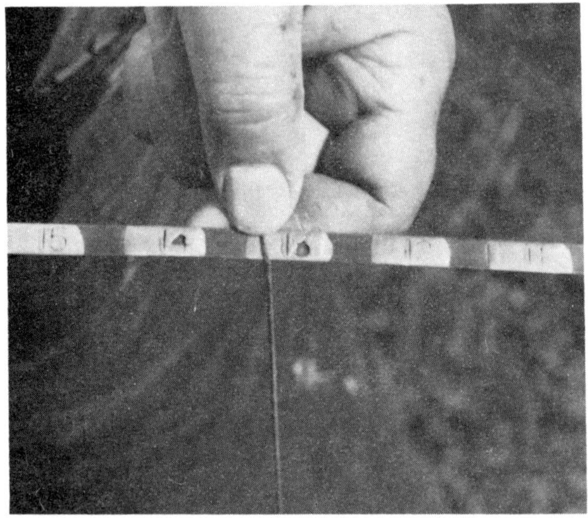

Figure 1-19. Manipulation of plumb bob string at zero end of tape.

The method just described is used when measuring the distance between two points, whether it be slope distance or horizontal distance.

To measure and set points on a line is difficult and should not be attempted by the beginning surveyor. Take for example a line whose end terminal is visible from the beginning point but at different elevation. Irrespective of whether horizontal or slope distance between the two end points is needed, extreme care has to be taken to make certain that every point set for measuring is on line. Sometimes this is not possible because of the terrain. If the length of the line exceeds the length of the tape a measurement of the slope must be made for each slope distance chained in order to reduce the slope distance to horizontal distance. It would be better to run a random line, setting stations at intervals less than the length of the tape being used. The length of the whole line and difference in elevation could then be computed trigonometrically.

Accurate chaining in rough to very rough terrain requires extreme care on the part of the chainmen. Most surveying done in forestry does not merit time spent on the study of theory of probability, probable error, standard deviation, etc. This is not to say that they should not be a part of surveying but they have no place in most kinds of surveying done in forestry.

Most errors in surveying are the result of poor chaining but it is next to impossible to eliminate or compensate for all possible errors. In most circumstances, the best accuracy that can be expected in chaining is 1:3000. A variation of 5 lbs. in tension will introduce an error of 1:10,000 in a 100 foot tape supported at the ends. Thus this error is going to increase when a 200 foot tape is used supported at the ends and there is considerable difference in elevation between the ends of the tape.

Corrections can be made for sag, tension and temperature but generally these corrections are not applicable to forest surveying. Most tapes are calibrated when supported throughout the entire length of the tape at a standard tension and temperature. It is apparent that these conditions are rarely encountered in the field and only in rare cases are corrections applied for sag, tension and temperature.

If all of these factors were considered, the cost of surveying in rough terrain would be astronomical. Shown below are approximate errors possible in using a 100 foot tape.

1. Length of tape not standard $\pm$ 0.01 feet

2. Tape not pulled tight (center 0.7
    feet out of line) $-$ 0.01 feet

3. Temperature other than $68°$ (for
    every $15°$ difference) $\pm$ 0.01 feet

4. Sag (middle of tape 0.6 feet below
    ends of tape) $-$ 0.01 feet

5. Tension (for every 15 pounds of
    pull too great) $-$ 0.01 feet

When one considers that it is very improbable that the chainman, using a plumb bob or when chaining on steep slopes can measure more accurately than the nearest $\pm$ 0.05 feet, it is impractical to consider corrections for other possible deviations from standards for which the tape was standardized. The best advice is to apply plenty of tension when chaining and be as careful as one can be in observing the recording data.

ANGLES

Since distances and angles measured in the horizontal plane are used together in all surveying problems, it is necessary to fully understand bearings and azimuth.

True bearings of lines or courses are the angles which the line of sight makes with the meridian passing through one end of the line or course. The angles, which are measured in degrees, minutes and seconds are measured east or west of the north-south line but never north or south of the east-west line. No bearing can exceed 90° since the circle is divided into 4 quadrants and the directions are given as being N-E, N-W, S-W or S-E. In Figure 2-1, the true bearing of the lines are as follows:

Bearing OA = N 60°E   Bearing OC = S 10°W
Bearing OB = S $80^\circ30'$E   Bearing OD = N $70^\circ45'$W

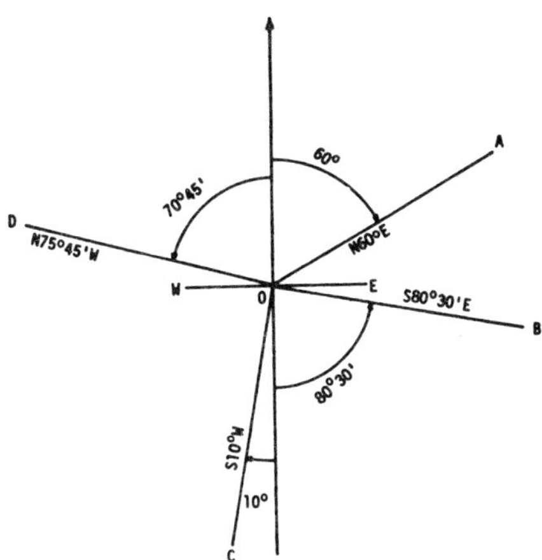

Figure 2-1. Bearing angles and quadrants.

The compass needle does not point toward true north except in certain areas but points to magnetic north by aligning itself in the magnetic lines of force in and around the surface of the earth. The angular difference between true north and magnetic north is known as the magnetic declination. The delination is East or E if magnetic north lies east of the true meridian (Figure 2-2) and West or W if magnetic north lies west of true north.

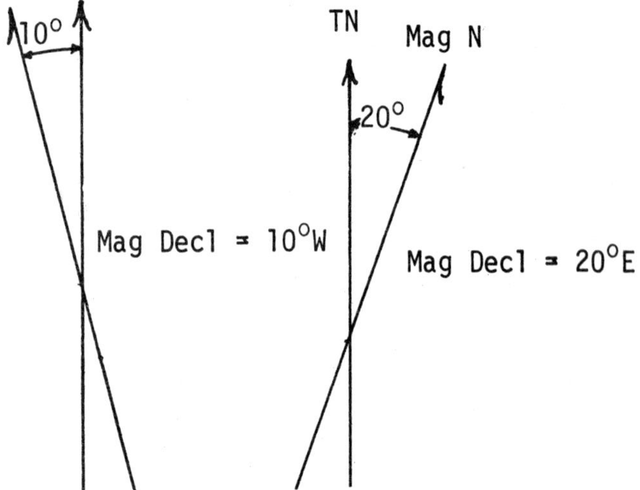

Figure 2-2. Magnetic declination.

All bearings read on the compass will be true bearings if the declination is set off on the compass by rotating the compass dial in the direction indicated by the declination in the area (Figures 2-3a, b, c).

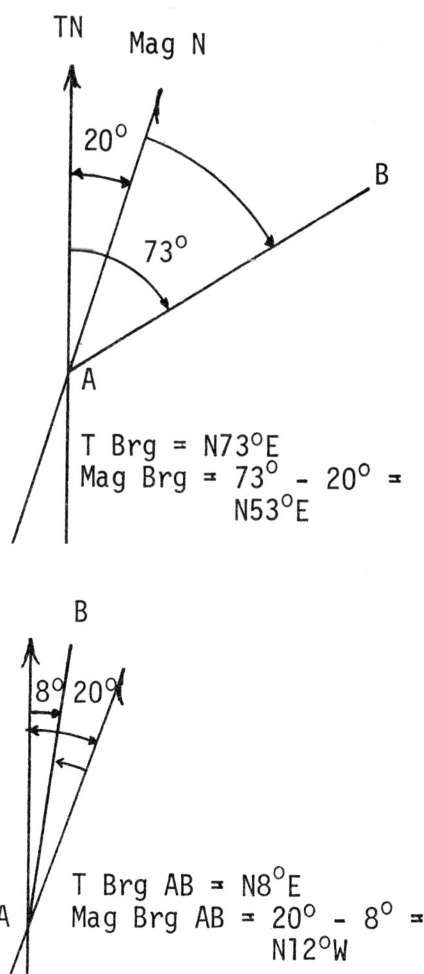

Figures 2-3a, b. Magnetic declination, true bearings and magnetic bearings.

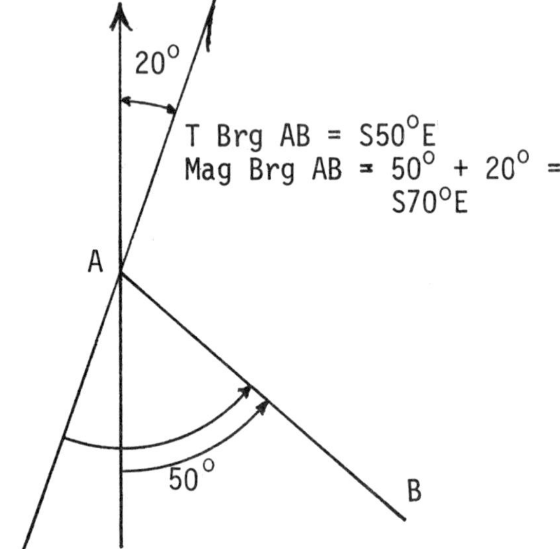

Figure 2-3c. Magnetic declination, true bearings and magnetic bearings.

If the declination is not set off on the compass, the compass will read the magnetic bearing of the line, which is defined as the angle which the line of sight makes with the magnetic north-south meridian. Hence, the difference between the true bearing and the magnetic bearing of a line is the magnetic declination (Figures 2-4a, b, c).

The direction of a line or course may also be expressed as an angle from 0 to 360° measured from true or magnetic north. These are known as true azimuth or magnetic azimuths respectively. The former is the most common method of expressing azimuth (Figures 2-5, 2-6). All bearings and azimuths are true unless stated otherwise.

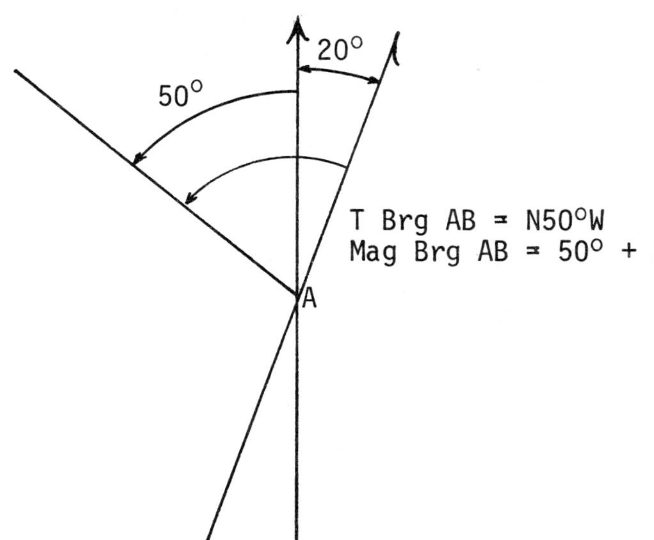

Figure 2-4a. Magnetic declination, true bearing and magnetic bearings.

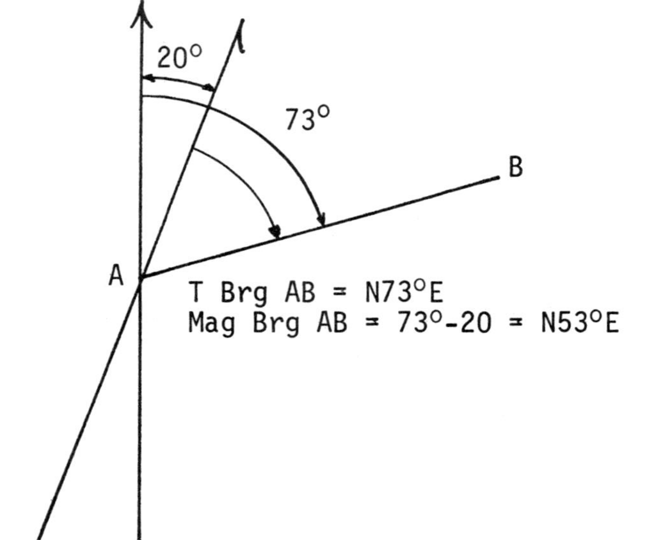

Figures 2-4b, c. Magnetic declination, true bearing and magnetic bearings.

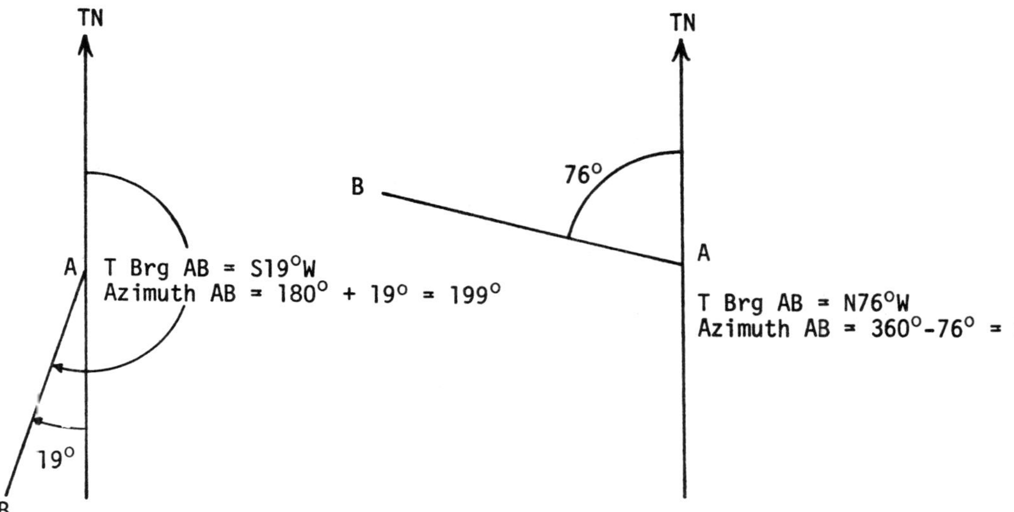

Figure 2-5. True bearings and azimuths.

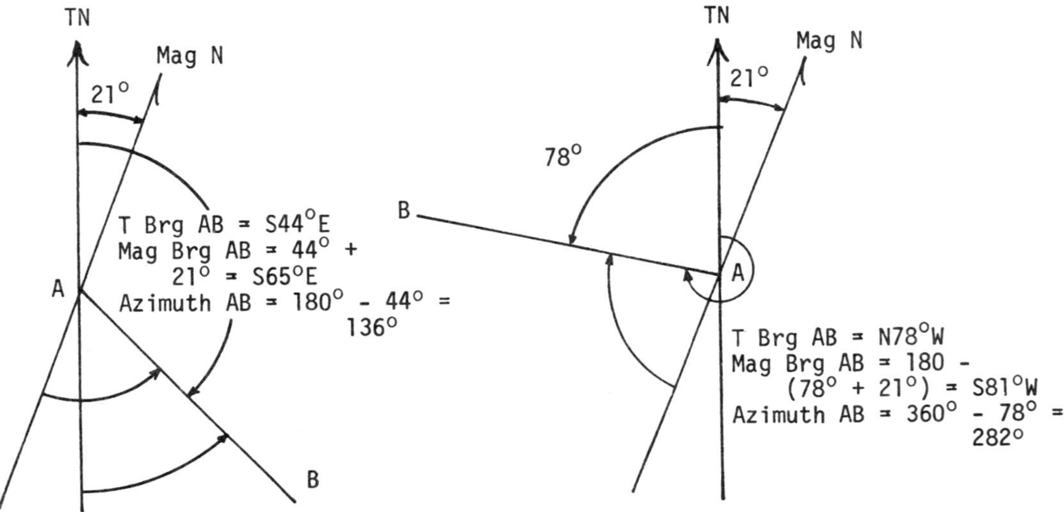

Figure 2-6. True bearing, azimuths and magnetic bearings.

Every line of definite length will have two bearings, one taken at each end of the line toward the other end. The bearing taken in the direction in which the survey is being made is known as the foresight (FS) and a bearing taken in the opposite direction is known as the backsight (BS). When recording the bearing of a line, the letter at the end of the line from which the bearing is taken is written first and the second letter is the point sighted on. Thus, there should be no problem determining the direction of a bearing taken. The bearing CD = N 36°30'E and is a foresight. The bearing CD = S 36°30'W is a backsight (Figure 2-7).

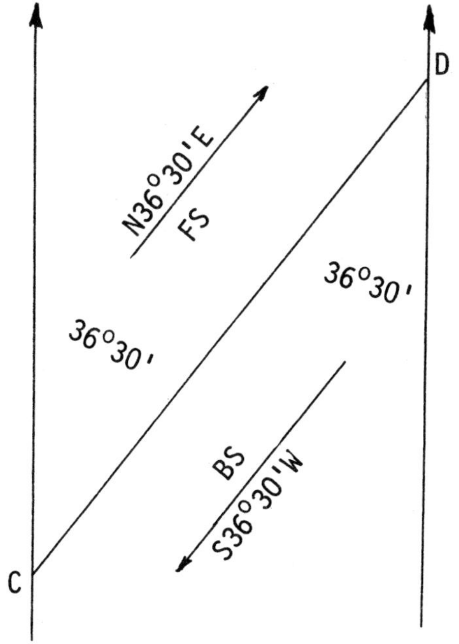

Figure 2-7. Foresights and backsights.

The backsight and foresight along a line should be numerically equal because alternate interior angles created by cutting two parallel lines with a transversal are equal but with a change of letters indicating the direction and quadrant of the sighting. If they are not equal numerically, it is due to location attraction. Local attraction is the result of the presence of certain types of minerals, power lines, fences, cables, etc.

Figure 2-8 illustrates an example of local attraction. The bearing from B to C is a foresight and is given as bearing BC = N 70°E. The bearing from C to B is a backsight bearing and is given as bearing CB = S 72°W. The bearing BA is a backsight bearing and = S 50°W. The bearing CD is a foresight bearing and = S 26°E. The bearing ED is a backsight from E to D and = S 20°W.

In the traverse shown in Figure 2-9, the following angles should be equal to each other because they are alternate interior angles measured from the meridian. The actual values determined by measuring the angles in the field with a staff compass were as follows and the problem is to determine the correct bearing of each course. The corrected bearing (corrected for local attraction) is always the foresight bearing.

∡ 1 = ∡ 2	∡ 5 = ∡ 6	∡ 9 = ∡ 10
∡ 3 = ∡ 4	∡ 7 = ∡ 8	∡ 11 = ∡ 12

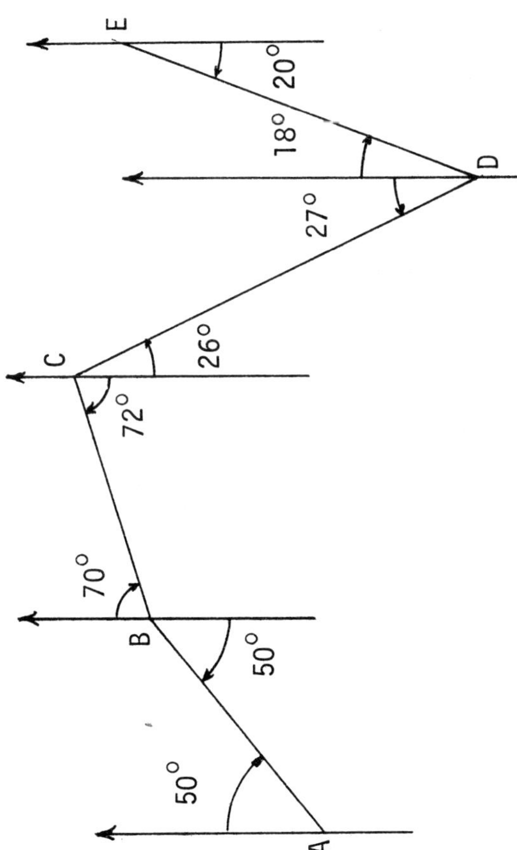

Figure 2-8. Foresights and backsights.

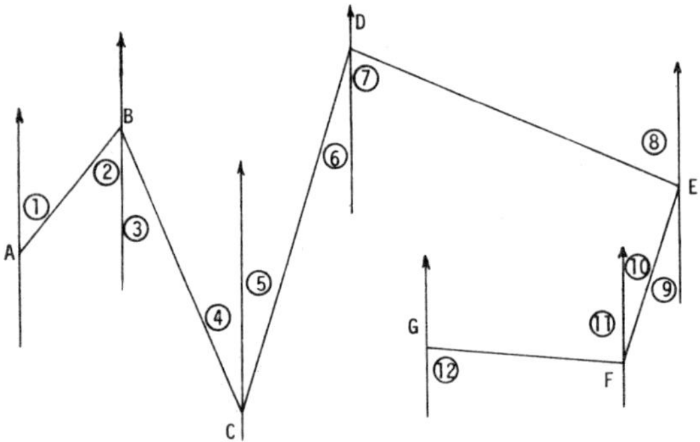

Figure 2-9. Correcting for local attraction in a traverse.

Station	Foresight	∡	Backsight	∡
G				
	N 83°45'W	11	S 83°00'E	12
F				
	S 10°15'W	9	N 11°00'E	10
E				
	S 67°15'E	7	N 65°00'W	8
D				
	N 17°30'E	5	S 18°00'W	6
C				
	S 22°00'E	3	N 21°00'W	4
B				
	N 40°00'E	1	S 40°00'W	2
A				

The first thing to look for in any traverse, where local attraction might exist, is a backsight and a foresight which are equal numerically. The fact that these two bearings (backsight and foresight) are equal indicates that there is no local attraction existing at the extremities of this line and any bearing taken at either end of the line will be correct in that there is no local attraction affecting the bearings. In this example, the foresight of AB and the backsight BA are equal to 40°. Hence no local attraction exists at A or B and the foresight bearing BC which equals S 22°00'E is correct because it was taken at station B.

The angle BCD is determined by using the backsight CB and the foresight CD. Both of these bearings are taken at station C and if there is any local attraction in this area, it will affect both bearings by the same amount.

$\angle$ BCD = $\angle$ 4 + $\angle$ 5 = N 21°W + N 17°30'E = 38°30'

—C— If there were no local attraction at station C, the backsight CB would have been numerically equal to the foresight BC which is S 22°00'E. This for the reason that $\angle$ 1 = $\angle$ 2. Therefore, if the backsight CB were N 22°00'W and the angle BCD which equals $\angle$ 4 + $\angle$ 5 = 38°30', then $\angle$ 5 = 38°30' – $\angle$ 4 = 38°30' – 22°00' = 16°30'. Since $\angle$ 5 is the foresight CD, the bearing of CD corrected for local attraction becomes N 16°30'E.

—D— Angle CDE = $\angle$ 6 + $\angle$ 7 = backsight DC + foresight DE = 18° + 67°15' = 85°15'. If there were no local attraction at D, the backsight DC would = foresight CD which was just computed to be N 16°30'E. Hence, $\angle$ 6 would = $\angle$ 5 which = 16°30' and $\angle$ 7 = foresight DE = 85°15' – 16°30' = 68°45' or the foresight DE = S 68°45'E.

—E— $\angle$ DEF + $\angle$ 8 + $\angle$ 9 = 180°. If there were no local attraction at E then $\angle$ 8 would equal $\angle$ 7 which = 68°45'. Since $\angle$ DEF is constant and $\angle$ 8 + $\angle$ 9 = 65°00' + 10°15', $\angle$ 9 = 75°15' – $\angle$ 8 = 75°15' – 68°45' = 6°30' = bearing EF = S 6°30'W.

—F— Would the corrected bearing of FG = N 88°15'W or S 88°15'W?

1. Course	Foresight	Backsight
AB	S 30°20'W	N 30°30'E
BC	S 67°00'E	N 66°00'W
CD	N 17°45'W	S 17°15'W
DE	N 62°30'E	S 63°00'W
EF	S 3°00'E	N 2°30'W

All bearings given are true bearings

Magnetic declination = 18°00' W

Solve for the following after correcting for local attraction.

Deflection angle at station B

Azimuth of FE

Magnetic bearing of CD

Magnetic azimuth of BA

Bearing of DE

2. The sum of the interior angles of a regular polygon equals 1800°. Find the value of a deflection angle if the figure is traversed in a counter clockwise direction.

3.   The magnetic bearing of a line forming part of a property survey in 1870 was recorded as N 46°30'W. At that time the magnetic declination was 2°30'E.   If the magnetic declination in 1980 is 3°00'W, what is the magnetic bearing of the line in 1980?

4. What is the magnetic declination if the true azimuth of a line is 15°15' and the magnetic azimuth of the same line is 356°30'?

5.

Station	Foresight	Backsight
E		
	N 1°00′E	S 1°30′E
D		
	N 88°30′E	N 89°00′W
C		
	S 55°45′W	N 56°30′E
B		
	N 82°00′E	S 82°00′W
A		

Declination = 20°30′E

All bearings given are true bearings

Find the following after correcting for local attraction.

Bearing DC

Azimuth DE

Deflection angle at C

Magnetic bearing CB

Bearing of AC if the distance AC equals distance BC

6.

Course	Magnetic Bearing	
AB	N 29°15′E	Compute the delection angles.
BC	N 67°10′E	Compute the delection angles.
CD	N 39°30′W	
DE	S 82°45′W	Assume that there is no local attraction.

7.

Station	Deflection angle	
B	23°15′L	Compute the bearing of the remaining courses if the bearing of course AB is N 38°47′E
C	12°20′L	
D	68°15′R	
E	6°35′R	

TRANSIT

The transit is one of the most versatile and useful instruments available to the surveyor. With the transit, he can lay off horizontal angles, measure direction, vertical angles, differences in elevation and distances when using the stadia hairlines and prolong lines. It is essential to understand the operation of a transit and to be able to obtain the required degree of accuracy with a minimum expenditure of time.

The transit consists of three main parts: (1) upper plate, (2) lower plate and (3) leveling head. The upper plate consists of the telescope, vertical circle and vernier, horizontal axis, standards, level vials, 2 verniers (horizontal) and an upper motion **clamp** screw and tangent screw. The lower plate is a horizontal circular plate graduated on its upper side. The leveling head includes the lower motion clamp screw and tangent screw, leveling screws, ball and socket joint and the sliding plate.

The upper and lower plates are attached to the inner vertical and outer vertical spindles respectively. The axis of rotation is at the geometric center of the outer graduated horizontal circle. The outer spindle is seated in the leveling head which consists of the ball and socket joint and the sliding plate. The plumb bob hook is attached to the end of the spindle. Four leveling screws are threaded into the leveling head and are used to level the two plate levels set at right angles to each other on the upper motion. There are clamp screws and tangent screws for the telescope and for the upper and lower motions. When the telescope level is on the underside of the telescope, the instrument is said to be in the normal position. When the level tube is above the telescope it is said that the telescope is plunged or in the inverted position. Recognizing these positions is important when doubling deflection angles.

A conventional engineer's transit is shown in Figure 3-1. Figure 3-2 shows the cross section of a transit The part names for the cross section view shown in Figure 3-2 are listed by number on page 45 (Courtesy of Keuffel & Esser Co.).

The verniers are devices for reading the angular value to a unit that is smaller than the smallest graduation on the circle. When the graduated circle is divided into 30 minute units, the vernier will consist of 30 divisions on either side of the zero mark, each division being 1/30 smaller than the division on the graduated circle. The 30 divisions of the vernier are equal to 29 divisions on the circle. The smallest unit or size of one unit on the circle and the number of divisions on the vernier should be determined carefully before beginning work with any transit. The least count or smallest value of an angle that can be read is S/N where S = the value of one division on the circle and N = the number of divisions on the vernier.

Figure 3-3a illustrates an instrument which has a vernier with 30 divisions and a circle whose smallest unit of graduation is 30'. Transits that have the degree marks divided in 30' units are usually read on to the nearest minute when S/N = 30/30 = 1'.

Some degree graduations on the circle are divided into 3 equal parts which make each graduation equal to 20'. The vernier on these instruments will have 40 divisions on either

Figure 3-1. Transit (courtesy of Keuffel & Esser Co.).

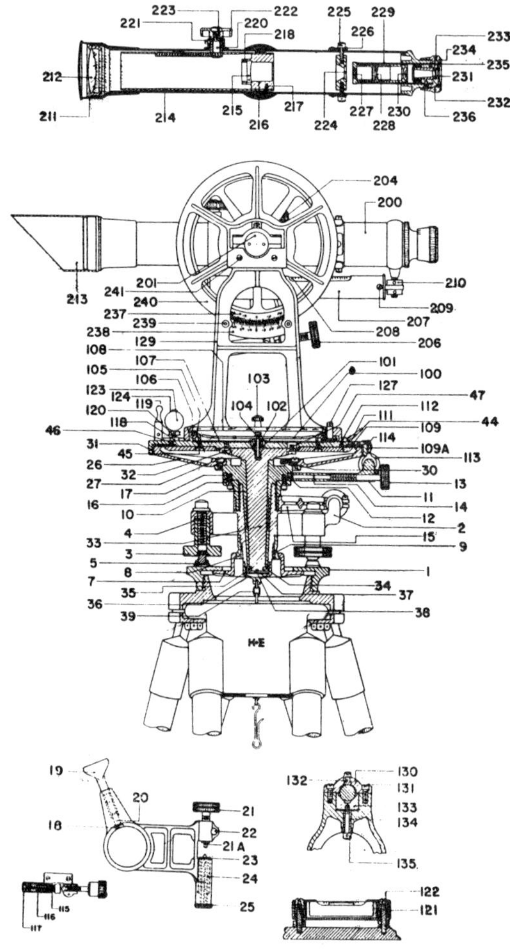

Figure 3-2. Cross section of transit and nomenclature (courtesy of Keuffel & Esser Co.).

1 Tripod Plate
2 Leveling Head
3 Leveling Screw Head
4 Leveling Screw Stem
5 Leveling Screw Shoe
6 Leveling Screw Cap
(Left Hand Thread)
7 Shifting Plate
8 Half Ball
9 Half Ball Lock Screw
10 Outer Center
11 Vernier Plate Clamp
12 Vernier Plate Clamp
Gib
13 Vern. Pl. Cl. Screw
14 Vernier Plate Clamp
Screw Pin
15 Lower Clamp
16 Lower Clamp Collar
17 Lower Clamp Collar
Mounting Screw
18 Lower Clamp Gib
19 Lower Clamp Screw
20 Lower Clamp Screw
Pin
21 Lower Clamp Tangent
Screw
21A Clamp Tangent Screw
Pivot Pin
22 Clamp Tangent Screw
Tension Screw
23 Lower Clamp Tangent
Screw Plunger
24 Lower Clamp Tangent
Screw Spring
25 Lower Clamp Tangent
Screw Cap
26 Horizontal Circle
27 Horizontal Circle
Mounting Screw
28 Horiz. Circle Vernier
29 Horiz. Circle Vernier
Mounting Screw
30 Horizontal Circle
Adjusting Screw
31 Vernier Plate
32 Vernier Plate Mounting
Screw
33 Inner Center
34 Center Nut
35 Center Nut Lock Screw
36 Center Cap
37 Center Spring
38 Center Ball
39 Plumb Bob Chain and
Hook
100 Compass Needle
101 Compass Needle Lifter
102 Compass Needle Lifter
Bushing
103 Compass Needle Lifter
Screw Assembly
104 Compass Needle Pivot
105 Compass Ring

106 Compass Ring Spring
107 Compass Dial
108 Compass Cover Glass
and Mount
109 Vernier Cover Glass
110 Vernier Cover Glass
Strap
110A Vernier Cover Glass
Strap Screw (not
shown)
111 Vernier Reflector
112 Vern. Reflector Frame
and Hinge
113 Vernier Plate Clamp
Tangent Screw
114 Vern. Pl. Cl. Tangent
Screw Spring Box
115† Tangent Screw Plunger
116† Tangent Screw Spring
117† Tangent Screw Cap
†(Not shown but similar to
Nos. 23, 24 & 25.) (For
Vernier Plate and Tele-
scope Cl. Tang. Screws)
118 Plate Level Bracket and
Posts complete
119 Plate Level Vial Guard
120 Plate and Standard
Level Adj. Nut
121 Plate and Standard
Level Spring
122 Plate and Standard
Level Post Cap
123 Plate and Standard
Level Vial, Tube and
Ends complete
124 Plate and Standard
Level Vial
125† Standard Level Post,
Fixed
126† Standard Level Post,
Adjustable
†Not shown.
127 Declination Adjustm'nt
Pinion and Washer
129 Standard
130 Trunnion Cap
131 Trunnion Cap Screw
132 Trunnion Friction Sc.
133 Trunnion Bearing Block
134 Trunnion Bearing Block
Adjusting Screw
135 Trunnion Bearing Block
Lock Screw
200 Tele. Barrel and Axle
201 Tele. Axle End Cap
202† Telescope Clamp
203† Telescope Clamp Gib
†Not shown.
204 Tele. Clamp Screw
205 Tele. Clamp Screw Pin
(Not shown).
206 Telescope Clamp
Tangent Screw

207 Tele. Level Vial, Tube
and Ends complete
208 Tele. Level Vial only
209 Telescope Level Tube
End Lock Screw
210 Tele. Level Adjust. Nut
211 Objective Cap
212 Objective Lens and
Mount
213 Sunshade
214 Telescope Draw Tube
215 Tele. Focusing Lens
216 Telescope Focusing
Lens Mount
217 Tele. Focusing Lens
Mount Lock Screw
218 Telescope Focusing
Lens Lock Ring
219 Tele. Focusing Pinion,
Pinion Head (222) &
Screw (223) complete
220 Tele. Focusing Pinion
221 Telescope Focusing
Pinion Lock Screw
222 Telescope Focusing
Pinion Head
223 Telescope Focusing
Pinion Head Screw
224 Reticule
225 Reticule Adjust. Screw
226 Reticule Adjusting
Screw Shutter
227* Eyepiece Lens I and
Mount
228* Eyepiece Lens II and
Mount
229* Eyepiece Tube
230* Eyepiece Lens III and
Mount
231* Eyepiece Focusing
Lens and Mount
232* Eyepiece Focusing Ring
233* Eyepiece Focusing
Ring Set Screw
234* Eyepiece Cap
235* Eyepiece Focus. Sleeve
236* Eyepiece Focusing
Sleeve Screw
237 Vertical (Stadia) Circle
238 Vertical Circle Vernier
239 Vertical Circle Vernier
Posts and Nuts
240 Vertical Circle Guard
241 Vert. Cir. Guard Screw

side of the zero mark. This type of instrument which is illustrated in Figure 3-3b would have a least count of S/N = 20/40 or ½′ or 30″.

**GRADUATED 30 MINUTES READING TO ONE MINUTE
DOUBLE DIRECT VERNIER**

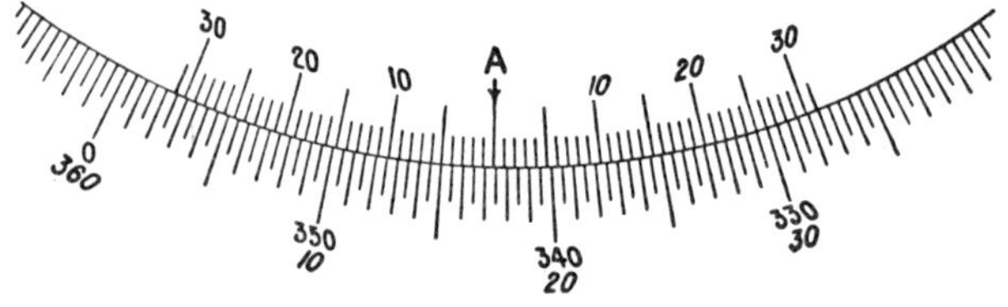

Figure 3-3a. One minute vernier.

**GRADUATED 20 MINUTES READING TO 30 SECONDS
DOUBLE DIRECT VERNIER**

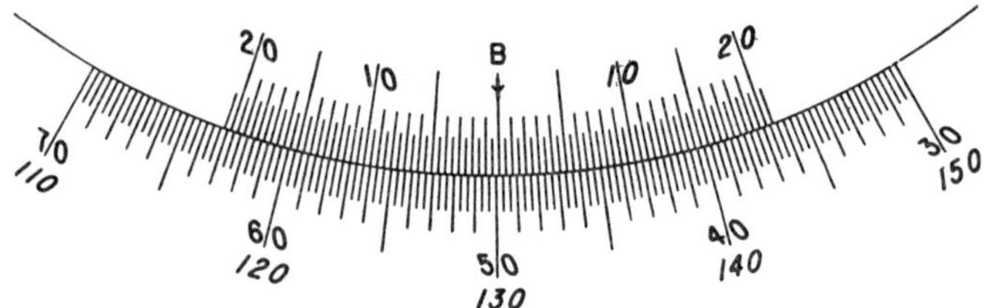

Figure 3-3b. Thirty second vernier.

Figure 3-3c shows an example of a scale and a vernier on which the angle could be read to 20″.

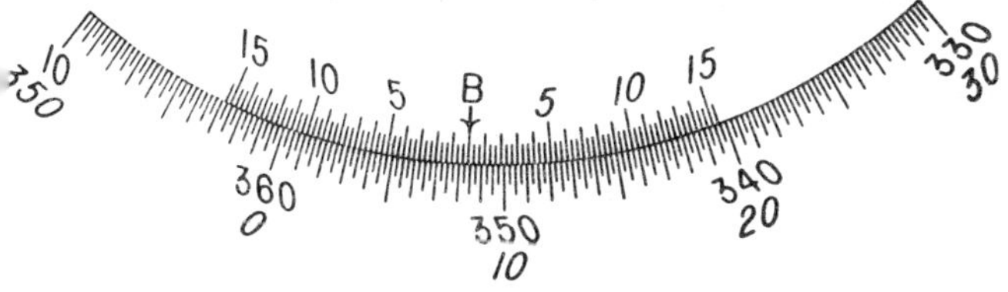

Figure 3-3c. Twenty second vernier.

Many errors are made in measuring horizontal angles with a transit by reading the wrong side of the vernier. Remember, always read the side of the vernier corresponding to the direction in which the vernier was moved. When measuring deflection angles, read the left side of the vernier if the line of sight is turned in a clockwise direction and read the right side of the vernier if the instrument has been turned counter clockwise. The sum of the readings taken on both sides of the vernier will total 30′. In Figure 3-4, if a reading of $14°17′$ were obtained, the zero mark of the vernier should be between $14°$ and $14°30′$. If the other side of the vernier had been read, a value of 13′ would have been obtained which would indicate that the zero mark of the vernier should be closer to the $14°$ mark than to the $14°30′$ mark. The correct reading therefore would be

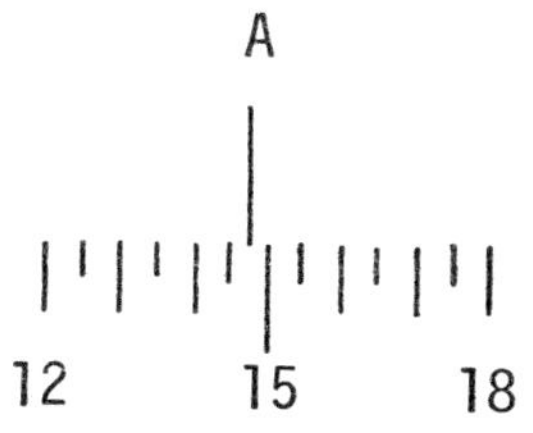

Figure 3-4. Sketch of vernier.

$14°30′$ plus 17′ or $14°47′$. Carefully noting the relationship between the zero mark on the vernier and the degree or 30′ mark on the circle should help eliminate this type of error. Failure to add the 30' to the value of the first angle read when doubling angles is also a very common source of error. Another error often made is to mistake the direction of movement of the vernier and read the angle as being 15°13'. Careful checking would indicate that the vernier was moving to the right and that the angle was less than 15°. Remember, that regardless of the direction of movement of the vernier, the angle is read in the direction in which the numbers are becoming larger.

On the vernier shown in Figure 3-5, there are eight possible readings that could be made and they would be as follows: 54°02'30", 54°17'30", 65°42'30", 65°17'30", 305°42'30", 305°17'30", 314°17'30" or 314°02'30". Only two of the values are correct depending on the direction in which the angle was turned. Which values are correct?

The circles on the instrument are not graduated perfectly and in some instances it is desirable to read both the A and the B verniers. However, if the horizontal angles are read with the telescope in both the normal and inverted positions when doubling angles, reading the A vernier twice is adequate. The A vernier is always located under the eye piece of the telescope when the telescope is in the normal position.

In setting the instrument up, two things must be accomplished: (1) "centering" it on the hub tack or point and (2) "leveling" to make the vertical axis or the transit's spindle truly vertical. This will automatically make the graduated circles lie in a horizontal plane.

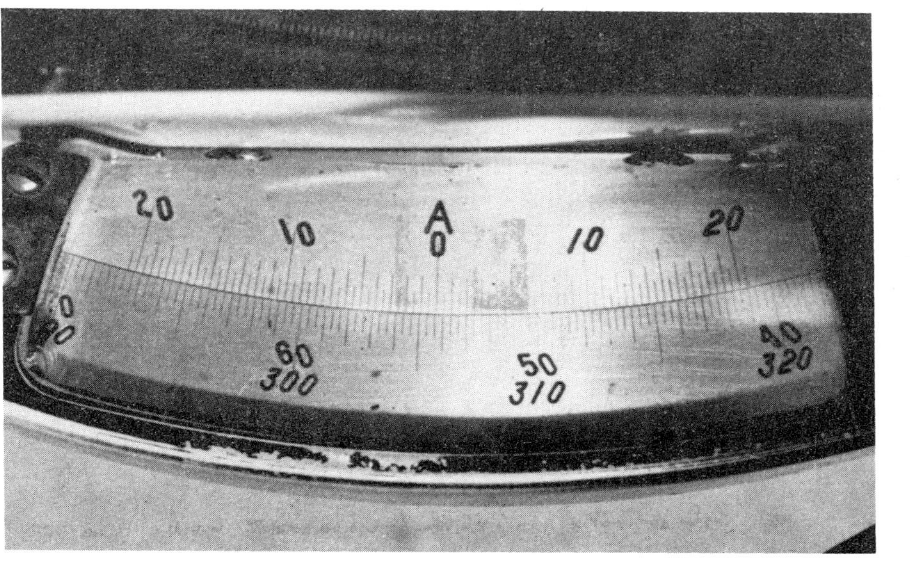

Figure 3-5. Enlargement of 30″ vernier.

More time is spent in setting up the transit than in any other operation while obtaining data in the field. Ability to set the instrument up quickly is worth a great deal because a slow transit man delays the work of the whole crew. Only practice will improve one's ability. The following are suggestions for quickly setting the transit up over a point.

1. Spread the tripod legs far enough apart to bring the telescope to a convenient height for sighting.

2. Manipulate the legs of the tripod so that the footplate is nearly level. This manipulation includes swinging the legs in and out or from side to side. On level ground the tripod legs are equally spaced from the hub tack. On a slope, place two legs on a contour line slightly lower than the point and place the third leg up the hill making sure that the foot plate is fairly level. When the three legs are set in the ground, the plumb bob point should be within a ½" of the vertical line through the hub tack. If it is off more than 2", pick the instrument up by grasping two legs and move it. This is easily done because the wing nuts on the screws where the legs are attached to the instrument should be tight enough to hold the legs out at a slight angle, similar to the position that they are in when the instrument is set up over the point.

3. If and when the plumb bob is within 1"-2" of the point, push the legs firmly into the ground. This should not change the level of the foot plate much. Make certain that the clamp screws on the extendable legs of the tripod are tightened. The plumb bob is quickly brought into a position over the hub tack by first getting the point of the plumb bob, the hub tack and one leg of the tripod in a

straight line. The pushing of a leg into the ground will cause the plumb bob to move toward that leg. Usually only two legs will have to be handled. Only as a last resort should the legs be lengthened or shortened (this will have the same effect as pushing the leg further into the ground) because it may loosen the other two legs or cause the foot plate to go off level. Figure 3-6 shows that by shortening or pushing leg 1 further into the ground, the plumb bob will move from position 0 to position I. Then by manipulating leg 2, the plumb bob will move to position II. The plumb bob should now be within a ½" (horizontal) of the hub tack and should be adjusted vertically until the point of the plumb bob just barely clears the hub tack.

4. The upper motion or lower motion clamp screws or both should be loosened in order to rotate the instrument about its vertical axis so the plate levels (set at right angles to each other) can be oriented with the axis of each plate level parallel to a plane passing through opposite leveling screws on the leveling head. By positioning the instrument in such a manner, only one level vial is effected by turning opposite leveling screws. The opposite leveling screws are turned in opposite directions at the same rate and the direction turned is determined by the position of the bubble in the level vial.

Level one plate level approximately, then repeat the process with the other level vial and the opposite pair of leveling screws. It is assumed now that the plumb bob is off of the hub tack approximately ½". Next turn all leveling screws enough to slightly loosen the screws on the foot plate. Now, without turning the foot plate, carefully slide the leveling head in the direction needed to center the

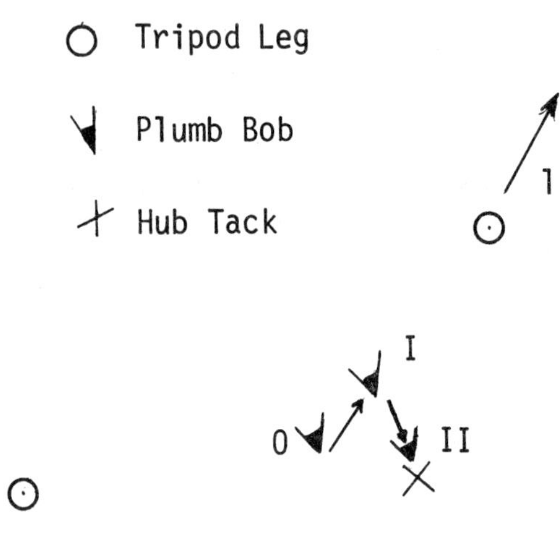

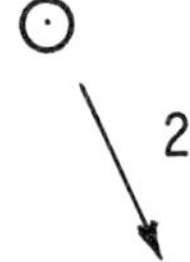

Figure 3-6. Manipulation of tripod legs to center plumb bob over hub tack.

plumb bob over the point, then tighten the instrument on the foot plate. If this is carefully done, the plumb bob will be centered and the instrument will be level. A small adjustment of the foot screws may still be needed to center the bubbles in the level vial. When the bubbles are centered, rotate the instrument 180° and check the bubbles to see that they are centered. If they are not centered the levels are out of adjustment, which means that the level vials do not lie in a plane exactly perpendicular to the vertical axis. Without adjusting the instrument, the bubbles are brought back toward the center of the level vial ½ of their original displacement using only one of the leveling screws. Turn the instrument 180° again and repeat the same process. The bubbles should remain in the same position as the instrument is rotated and the vertical axis is now vertical.

The first step taken to measure any horizontal angle with the transit is to set the A vernier on "0". With both the upper and lower motion clamp screws loosened, the lower plate is turned until the zero of the A vernier and the zero of the circle are nearly lined up. Tighten the upper motion clamp screw and with the aid of a hand lens, bring the zero marks into line with each other by turning the upper motion tangent screw. When the zero marks are lined up, the lines adjacent to "0" on the vernier should have equal offsets on the opposite scales. Care must be taken to make certain that you sight perpendicular to the horizontal circle when zeroing the vernier. The B vernier could also be checked but it is only done when a high degree of accuracy is needed such as in triangulation.

In order to measure any angle, whether it be a deflection angle, an angle to the left or an angle to the right it is necessary to start from a backsight. The upper motion clamp should be tight and the vernier set on zero. The lower motion clamp screw is loosened and the instrument is aimed in the direction of the backsight station by looking over the top of the telescope. When the point is in the field of vision, the lower motion clamp is tightened and the telescope is moved up and down until the backsight point is approximately in the center of the field of view. Fine adjustment is made with the lower motion tangent screw and the telescope tangent screw. Be certain that the hairlines are clear and distinct and that no parallax exist. Move the eye very slightly from side to side while sighting through the telescope, which, if parallax is present causes the vertical hairline to appear to move back and forth across the point. This is the result of the object being focused at some point not quite in the plane of the reticle ring which contains the hairlines. Proper focus of the objective and hairlines will eliminate the parallax.

If an angle is to be turned from the backsight, the upper motion clamp screw is loosened and the telescope rotated to point toward the forward station. Sight over the top of the telescope to line up on the point and then look through the telescope to see if the point is in the field of view. If it is, tighten the upper motion clamp and by using the upper motion tangent screw, bring the hairline onto the point on which the foresight is being taken.

If a deflection angle is to be measured, it is recommended that the telescope be inverted prior to loosening the upper motion clamp screw so the instrument man standing behind the transit may be certain of the direction that the deflection angle is being turned, right or left, especially if it is a small angle.

For most purposes of surveying in Forestry, the doubling of deflection angles is preferred for the following reasons.

(1) Angles are read with the telescope in both the normal and inverted position, giving average values if the vertical hairline is slightly out of adjustment.

(2) Doubling angles eliminates angular error which might occur due to improper backsights or foresights, vernier not set on zero to begin with or not reading the angle values correctly. It is the quickest means of obtaining the accuracy desired for 1/5000 closure.

Figure 3-7 is a plan view of a transit at station B taking a backsight on station A with the telescope in the normal position.

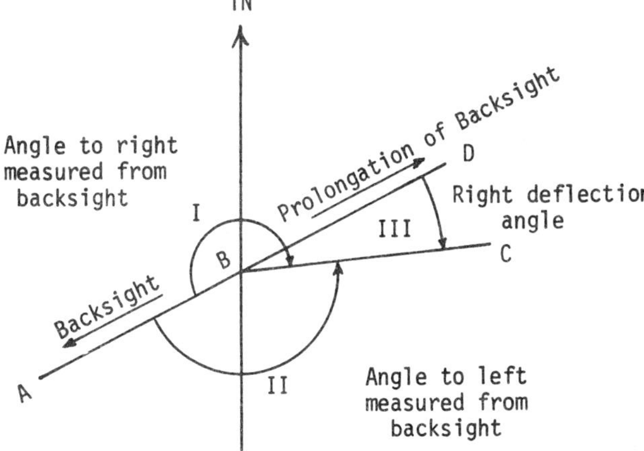

Figure 3-7. Drawing of angle right, angle left and deflection angle.

56

∡ I is an angle to the right, measured to the right from the backsight line. The angle may be measured more than once but the telescope usually remains in the normal position. The angle may be an interior angle or an exterior angle and may be less than or greater than $180°$.

∡ II is an angle to the left measured from the backsight line.

∡ III is a right deflection angle. Line BD is the prolongation of the backsight line BA. Prolonging the line is done by inverting or plunging the telescope. From BD a right deflection angle is turned to sight on C. An angle turned in a counter clockwise direction from BD would be a left deflection angle.

If the azimuth of the line AB were $65°$, the azimuth of BA would be $245°$. By setting the A vernier on $245°$ (back azimuth) for the backsight, the telescope could be turned clockwise or counter clockwise to any point without plunging the telescope and the azimuth of the new sighting would be read directly from the A vernier. This would eliminate the calculation of the bearing or azimuth of the foresight, which would be required if an angle to the right or left or if a deflection angle were measured. The shortcoming of this method is that the angle is turned only once and there is no way of checking for errors.

Following are some common errors made in measuring horizontal angles:

1. Loosening the wrong clamp screw.

2. Failing to tighten the clamp screws.

3. Turning the wrong tangent screws.

4. Reading the number on the horizontal scale from the wrong scale.

5. Reading the angles in the wrong direction.

6. Reading the vernier in the wrong direction.

7. Dropping 30'.

8. Reading the wrong vernier.

Vertical angles are read on the vertical circle with the aid of the vertical vernier. The full vertical circle, graduated into quadrants ($0°$-$90°$) is attached to the horizontal axis and turns as the telescope is raised or lowered while the vernier is attached to the standard and is stationary with respect to the circle. The vertical circle should read zero when the telescope is exactly level. The vertical angle is measured as being above or below the horizontal plane passing through the horizontal axis of the telescope. If the line of sight is above the horizontal plane it is said to be a + (plus) angle and if it is below the horizontal plane it is called a – (negative) vertical angle. The vertical vernier operates on the same principle as does the horizontal vernier. Since the vernier itself is fixed, it is read on the side away from the zero on the vertical circle, i.e., in the direction in which the vertical angle is **increasing**. Be certain to read the angle itself on the vertical circle. Figure 3-8 shows a reading of $-4°02'$.

The index error should be checked regularly. This is the error in the vertical angle due to a lack of parallelism between the line of sight and the axis of the telescope level tube. This error is determined by very carefully centering the bubble in the telescope level tube, after leveling the instrument with the plate levels, and reading the vertical vernier. The correction needed is equal to the index error but opposite in sign. For example: if the index error were found to be –02′ and the vertical angle observed was +4°13′. the correct angle would be 4°13′ + 02′ or 4°15′. All vertical angles should be double sighted, which means reading the vertical angle once in the normal position and once in the inverted position and taking an average of these two readings. The two angles read should not differ by more than 04′.

Figure 3-8. Vertical circle and vernier.

The middle hairline is used to measure the vertical angle; care must be taken that the lower or upper hairline is not used. If an angle is measured using the middle hairline the first time and either the upper or lower hairline the second time, a discrepancy of 15'-18' will exist between the value of the angles read. This large error would require a re-check of the sighting since one reading was obviously taken using the wrong hairline. **Warning** The vertical angle will double if any of the three horizontal hairlines are used twice, but reading the upper or the lower hairline will result in an error in the vertical control of a traverse that is very difficult to find.

Another method of measuring vertical angles employed in some instruments is to measure a zenith angle; i.e., from a point directly over the plumb bob down to the horizon, the point directly overhead being $0°$. An angle of $90°$ or $270°$ thus reads to the horizontal plane and is the same as a $0°$ reading on the vertical circle of a transit.

THEODOLITES

An optical reading theodolite is really a transit of high precision. Most theodolites in use are repeating theodolites with erecting type eye pieces. A direction theodolite is a non-repeating type of instrument which does not have a lower motion. A repeating theodolite is one on which successive measurements of horizontal angles may be accumulated. The graduated horizontal and vertical circles are made of glass and viewed through a sight adjacent to the telescope by means of prisms and mirrors. Using this system, the vernier of the conventional transit has been eliminated.

Figure 3-9 shows the Lietz T-60D theodolite and the instrument nomenclature. Figure 3-10 is a Kern K1-S Engineers theodolite and it too is a direct reading theodolite.

The theodolite is set up much as is a transit. It is attached to the tripod by means of a screw device and is roughly positioned over the point by using a plumb bob or a centering rod in the case of a Kern theodolite. After a rough leveling of the instrument using the bull's eye bubble, the theodolite is more exactly position over the point by using the optical plummet. The image of the ring mark of the optical plummet is sharpened by rotating the eyepiece and focusing is done by pushing in or pulling out the eyepiece. Exact leveling is done with the plate level, by orienting the plate level parallel to a line through two foot screws, and centering the bubble by using both foot screws. The instrument is then rotated 90°and the bubble is again centered using the third foot screw as is shown in the first two steps in Figure 3-11. When sights are steeply inclined

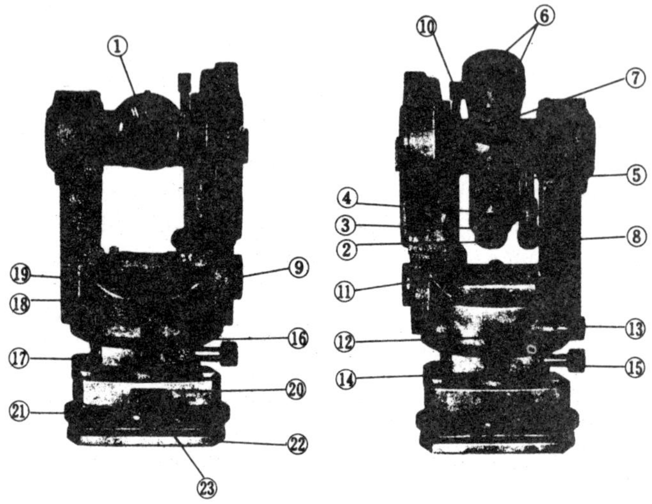

①	Telescope objective lens	⑬	Upper motion fine adjust screw
②	Telescope eyepiece	⑭	Lower motion clamp
③	Reticle focus ring	⑮	Lower motion fine adjust screw
④	Reticle adjustment cover	⑯	Circle positioning ring
⑤	Telescope focus ring	⑰	Circular level vial
⑥	Pointer sights	⑱	Optical plummet
⑦	Reticle illumination adjust knob	⑲	Optical plummet adjust screw
⑧	Circle reading eyepiece	⑳	Tribrach
⑨	Light reflector	㉑	Leveling screw
⑩	Telescope clamp	㉒	Bottom plate
⑪	Telescope fine adjust screw	㉓	Tribrach clamp & set screw
⑫	Upper motion clamp		

Figure 3-9. Theodolite and instrument nomenclature (courtesy of Lietz).

Fig. 1 and 2   Kern K1-S engineer's theodolite

1	Horizontal clamp	10	Circle reading eyepiece
2	Focusing ring	11	Lighting mirror
3	Optical plummet	12	Vertical clamp
4	Telescope objective	13	Terminal for the power supply for electric lighting or the DM 500
5	Vertical slow-motion screw		
6	Repetition clamp (not shown)		
7	Horizontal slow-motion screw	14	Horizontal level
8	Finder collimator	15	Leveling knob
9	Telescope eyepiece		

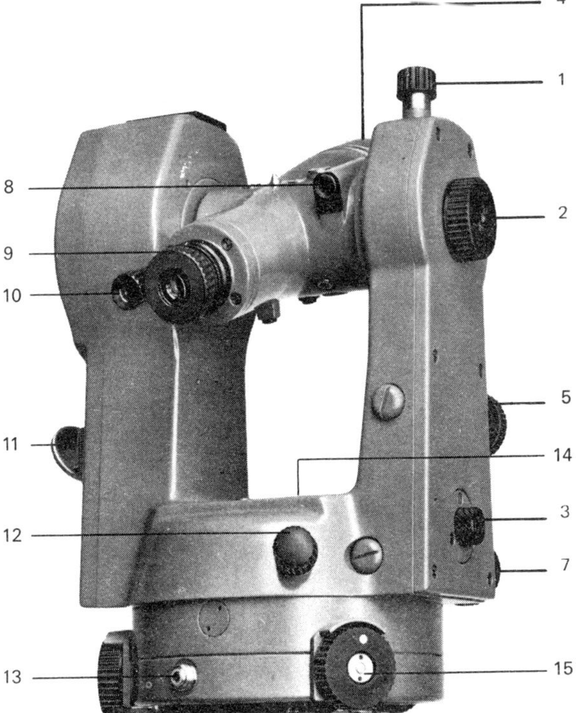

Figure 3-10 Kern K1-S Theodolite (courtesy of Kern and Co. Ltd).

leveling has great influence on the precision of horizontal direction measurement. Then it may be necessary to find the balancing point of the level by reversal. This is done by completing steps 3, 4, 5 and 6 in Figure 3-11.

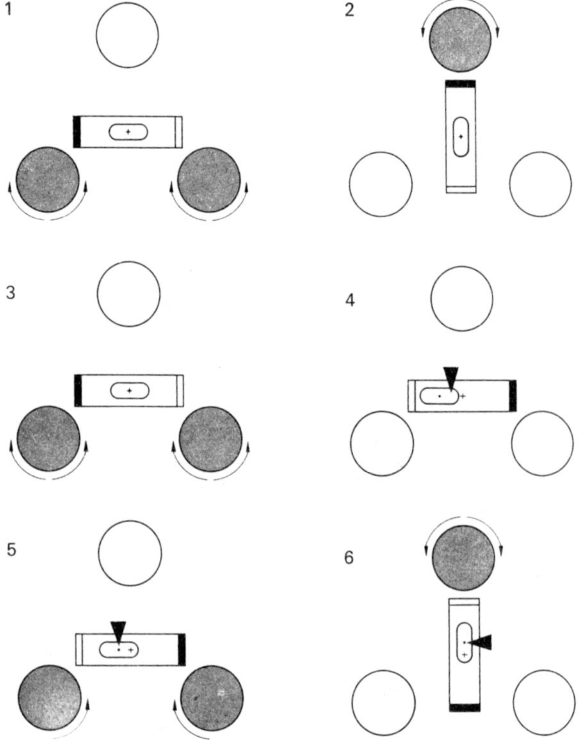

Figure 3-11. Leveling vial and foot screws of a Theodolite. (Courtesy of Kern & Co., Ltd.)

The horizontal and vertical circles are both read in the circle reading eyepiece after rotating and tilting the lighting mirror until the images are brightly illuminated. Figure 3-12 illustrates how the light is reflected on the scales in a cut away view of a Kern K1-S.

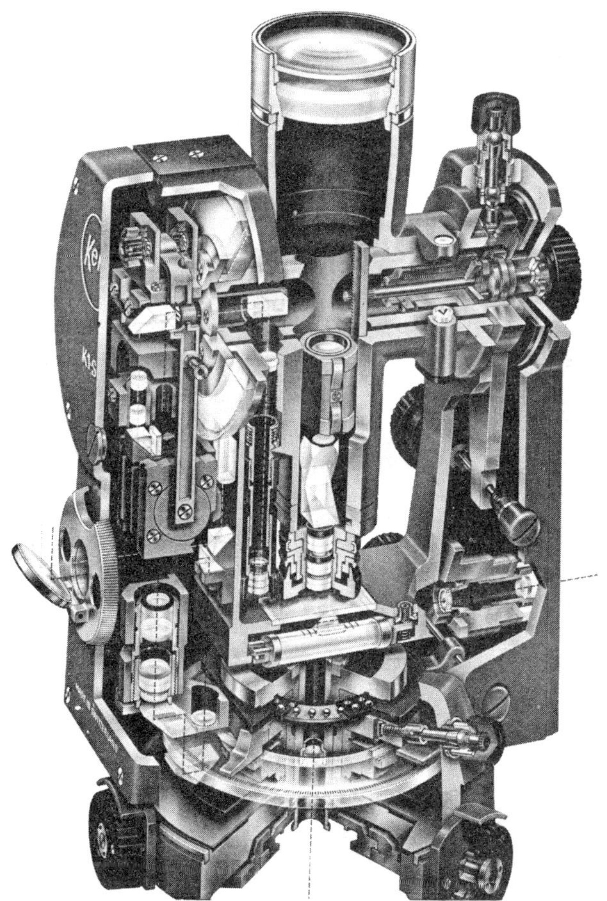

Figure 3-12 — Cut away view of K1-S Theodolite (courtesy of Kern & Co. Ltd.).

The scale to use when reading the horizontal angle is determined by the direction turned. This is aided by the small arrows in Figure 3-13a. The horizontal scales on the Kern K1-S are separated as illustrated in Figure 3-13b. Clockwise angles are read on the center panel and counter-clockwise angles are read on the lower scale as is indicated by the arrow pointing to the left. The vertical circle readings are zenith angles mean indicates that zero degrees is a point directly overhead.

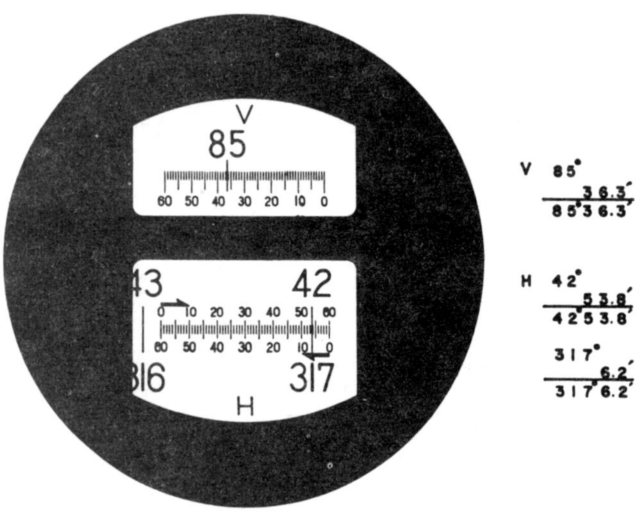

Figure 3-13a. Vertical and horizontal optical scales on a T-60D Theodolite (Courtesy of Lietz).

The optical scales illustrated are easy to read. The whole degree is obtained from the large numbered graduations and the minutes are read directly from the minute scale where it is cut by the index line. Tenths of minutes can be estimated with a little practice and the angle can be read to 6″ without difficulty.

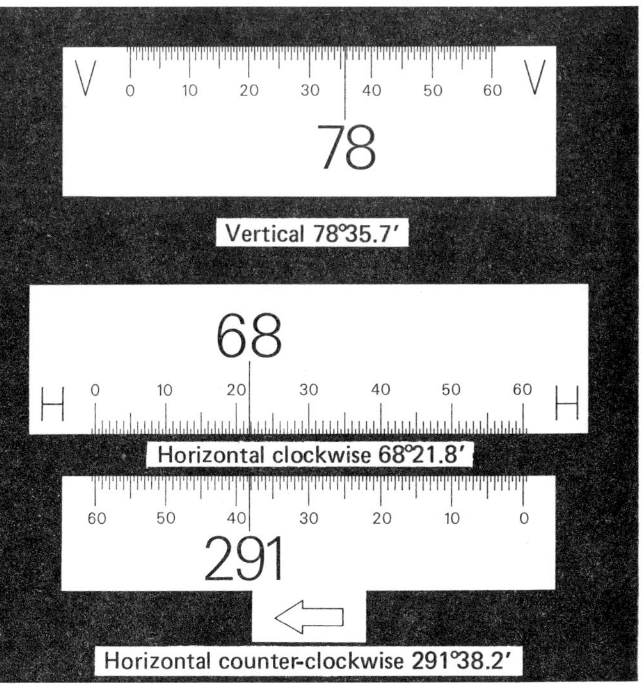

Figure 3-13b. Scales on K1-S Engineer's Theodolite
(Courtesy of Kern & Co., Ltd.)

TRAVERSING

Traverses are of two types: (1) closed traverses which return to the point of beginning and (2) open traverses, which start at one point and end at another. In either type the length of each course is measured and angles at one extremity of each course are measured. In most traverses the distance measured is the slope distance and the angles measured are either zenith or vertical angles and horizontal angles such as deflection angles. Figure 3-14 illustrates how the slope distance is reduced to horizontal distance and difference in elevation trigonometrically using the vertical angle. Figure 3-15 shows the method of converting horizontal distance to latitudes and departures, having used the deflection angles to compute the bearing on each line. The coordinates of each point are then computed for plotting purposes.

An example of a closed traverse would be one run around a tract of land and of an open traverse one run for a "P" (preliminary) line or route survey for road location. The error of closing a closed traverse can be checked mathematically but not so with the open traverse.

Discussed earlier were several methods of measuring horizontal angles at the end of a line. These terms are also applied to the type of traverse, i.e., interior angle, exterior angle, angles to the left or right and deflection angle traverse. The latter is the most common type used in forest engineering and is used to overcome the disadvantage of the other types of traverses; i.e., in most of them the angle is measured only once and even if it is doubled it is usually doubled with the telescope in the same position for both

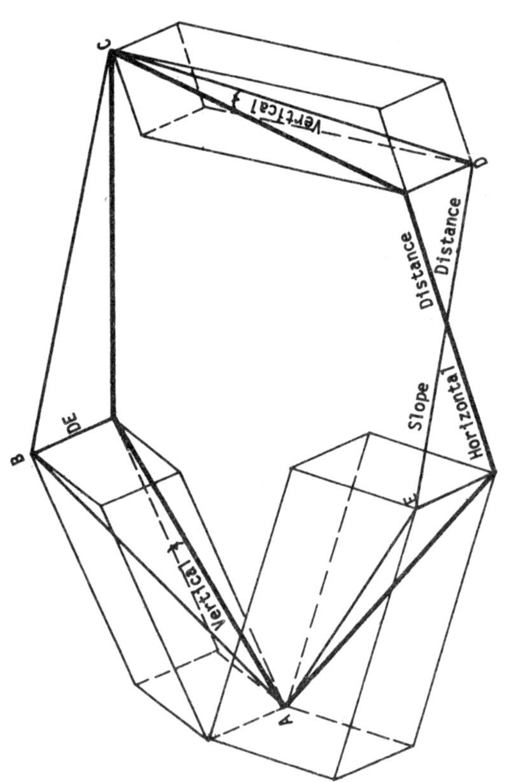

Figure 3-14. Slope distance reduction

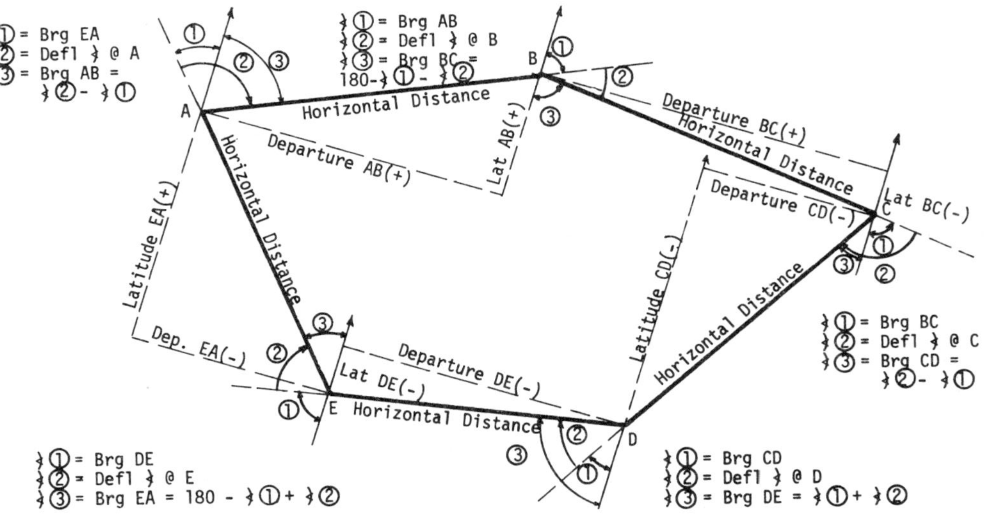

Figure 3-15. Reduction of horizontal distance to latitude and departure.

sightings. This does not compensate for or eliminate an error caused by the line of sight not being perpendicular to the horizontal axis. Deflection angles are easily measured twice, once with the telescope normal and once inverted thus decreasing the chance of error and increasing the accuracy to the degree needed for control type surveys.

When measuring a deflection angle put the telescope in the 'normal' position and, after setting the "A" vernier on zero, a backsight is taken on the previous point using the **lower motion clamp** and **tangent screws** to align the instrument. The line of sight is prolonged by plunging the telescope and a foresight is taken on the next station using only the **upper motion clamp** and **tangent screws.** After reading the angle, the procedure is repeated, taking a second backsight on the first point using the lower motion again but leaving the telescope in the inverted position. The line of sight is again prolonged, which returns the telescope to the normal position and using only the upper motion the angle is measured again. Since the angle is now the sum of the first and the second reading, the second value of the angle read on the "A" vernier is the doubled angle. The mean value of the doubled angle is the value used in all calculations. The measurement is of sufficient accuracy if the mean value of the doubled angle is equal to the value of the first angle + or - 30". This means that the value of the two angles measured agree with each other within 01'. The vernier is reset on zero at each new station.

While an azimuth traverse is less accurate because the angle is measured only once it also takes less time because the azimuth of each line need not be computed. The azimuth of the new line is determined merely by reading the vernier. The azimuth of the first line should be known or assumed, for example, assume the azimuth of line AB = $40°$. The instrument is set up on station B with the back azimuth of $220°$ set on the "A" vernier. A backsight is taken on A and after loosening the upper motion and without plunging the telescope, a foresight is taken on C. The "A" vernier is read and recorded as the value of the azimuth of BC. Without changing the setting of the verniers, the instrument is set on C and a backsight is taken on B using the lower motion. Since the B vernier is now set on the back azimuth of CB, the angle read on the B vernier after sighting from C to D is the azimuth of CD. The A vernier is now reading the back azimuth of CD. If the reading of the A and B verniers is alternated, the azimuth of each of the following courses is easily determined.

Sometimes in traversing it is necessary to prolong a straight line. The most accurate method is to set the instrument on station B and take a backsight on station A. The telescope is then plunged and point C′ is set. This in effect is the same as turning a deflection angle of zero degrees. After setting point C′, the lower motion is loosened and a backsight is again taken at station A. The telescope is again plunged and point C″ is set. The distance between C′ and C″, if there is any, is bisected and station C is set at the midpoint between C′ and C″. C is the station set on the prolongation of line AB. This method is known as double centering. Figure 3-16 shows how this process is repeated until the line has been prolonged the required distance.

Occasionally it is necessary to run a straight line between two points. This is not the same as prolonging a straight line. There are three conditions under which a line might be run or established between two points.

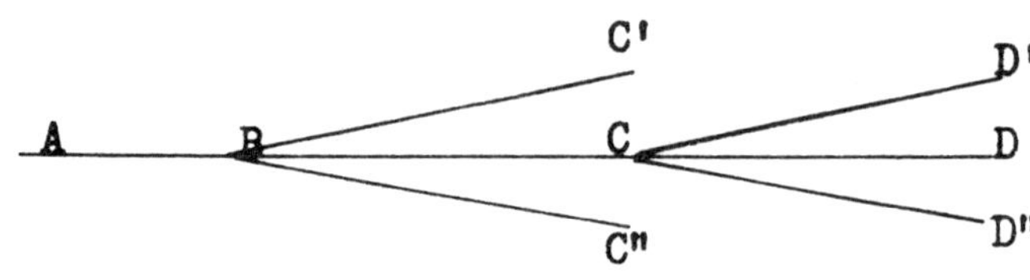

Figure 3-16. Prolongation of a straight line.

1. Terminals intervisible.

Points are set on line of sight after the instrument (transit or theodolite) is set on one extremity of the line and sighted on the other extremity.

2. Terminals not intervisible but are both visible from an intervening part of the line.

The instrument is set up nearly on line between the extremities. A backsight is taken on one end and the line of sight is prolonged by plunging the telescope. The position of the prolonged line of sight with respect to the point you are trying to sight is noted and the instrument is then shifted laterally to try to bring the prolonged line on target. Repeated attempts will finally bring it on. The amount to move the instrument can be computed rather than estimated provided the distances from the instrument to the ends of the line are known.

3. Terminals not visible from any intermediate point.

A random straight line from one terminal is run in the approximate direction of the far terminal, continuing until near the terminal point. A perpendicular offset from the terminal point to the random line is measured. Two solutions may now be followed: (1) Compute the angle between the desired line and the line and lay this angle off at the beginning terminal by repetition and extend this new line. (2) Compute the perpendicular offsets from the random line, and, having determined the angle between the random line and the desired line, set the points on the desired line using offsets.

A more practical method would be to run a random traverse from one terminal point to the other. After computing the bearing of the desired line and the coordinates of the points established on the random traverse, compute the coordinates of points on the desired line and make appropriate offsets to establish the desired points.

Irrespective of the type of traverse run, other detail is sometimes needed and plotted with respect to the stations established on the control traverse system. These details are obtained by side shots which may consist of measured distances, angles turned from sightings on the previous or following station, or a combination of both distance and angles. The dimensions of objects are usually measured although this is not always necessary. Figure 3-17 shows the location of detail by use of angles, distance and a combination of both.

Figure 3-18 illustrates the use of ties, range ties and swing offsets in locating detail.

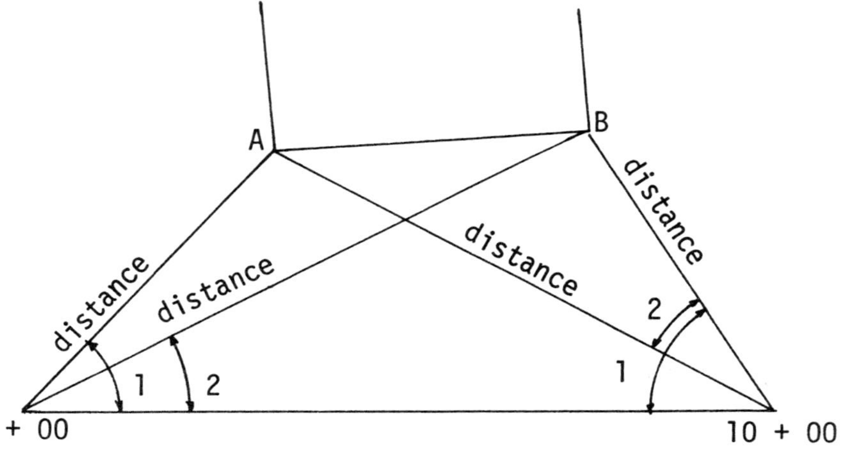

Figure 3-17. Intersection and radiation method of locating detail.

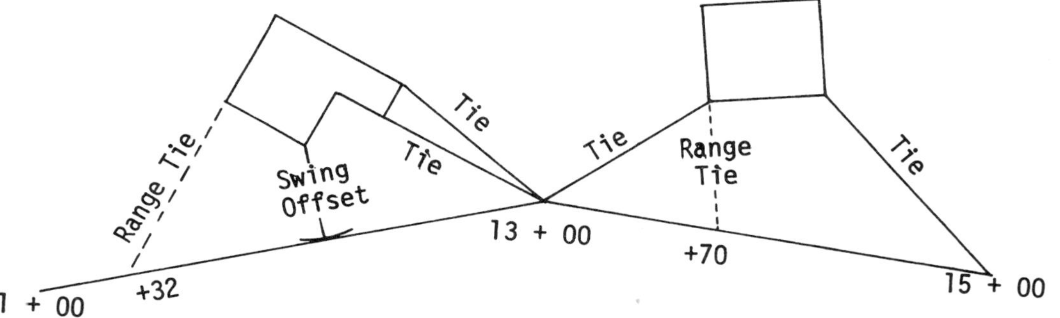

Figure 3-18. Range tie and swing offsets used to locate detail.

Range Ties — intersection of sight along side of building and traverse line.

Tie — distance from object to station.

Swing Offset — perpendicular distance from object to traverse line. Details of curved or irregular boundaries are also determined by making perpendicular offsets at regular intervals on a transit line.

CALCULATIONS

Before a closed traverse is plotted it is desirable to make certain calculations to determine the accuracy of the field data. Interior angles of a closed traverse should total $(n-2)(180°)$ where n is the number of sides. Deflection angles should total $360° \pm 30''\sqrt{n}$ where n again is the number of sides and right and left deflection angles are considered as of opposite sign. Should these checks not be met some field work may have to be repeated. Small angular errors can be proportioned around the traverse but there is little justification for making a correction to an angle when the correction is less than the smallest value which the instruments in use are capable of measuring. A closed traverse can be plotted before corrections are made, but if the plot does not close there is no way of knowing whether the data or the plotting is in error.

The deflection angles of an open traverse cannot be balanced as no geometrical requirement exists. The only check made is to measure vertical angles at each station of the traverse, then when the differences in elevation computed from the slope or horizontal distances and the vertical angles measured are summed, the sum should be within ± 1 foot of the DE between the beginning and end of the traverse. If this D.E. is not known extreme care in obtaining all field data is your only recourse.

Traverses are most easily and accurately plotted by the rectangular coordinate method. This is a rectangular grid system, the vertical or Y axis being the true north south meridian and the X axis an east-west line perpendicular thereto, i.e., a parallel of latitude. The intersection is the point of origin of the grid, and the position of any station on the grid is indicated by its total distance from each of these lines, the y coordinate being its distance N or S of the E-W reference line and its x coordinate the total distance E or W of the N-S reference line. Distances E and N are positive, S and W are negative. Values assigned to the origin are usually such that all points of the traverse will have positive x and y coordinates; i.e., the whole traverse will lie in the quadrant NE of the point of origin. Plotting a point by coordinates places that point only, its position is totally uninfluenced by the preceding and following points.

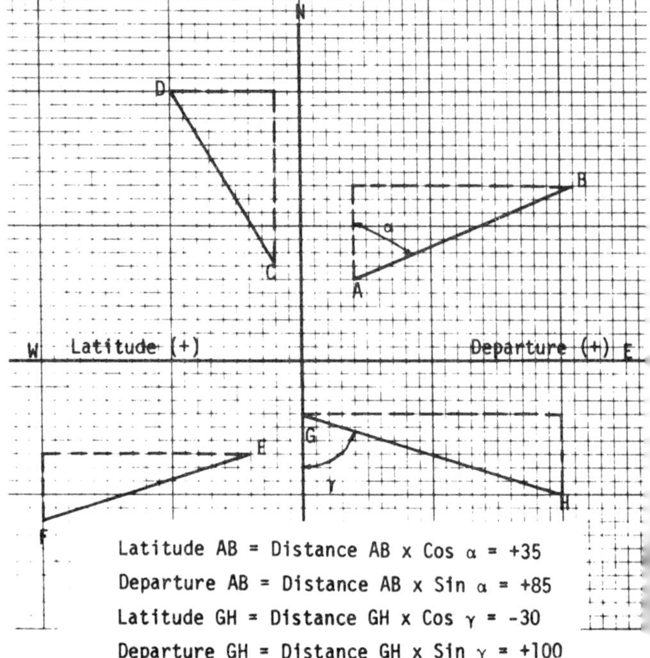

Figure 4-1. Latitudes and departures.

Coordinates at the intersection of X and Y axis = (0,0) i.e., Latitude Coordinate (Lat. Coord. or L.C.) = 0 and Departure Coordinate (Dep. Coord. or D.C.) = 0. The coordinates of point A are: L.C. = +60 and D.C. = +40 as determined along the Y and X axis respectively and the L.C. and D.C. of B are 130 and 210 respectively. The coordinates of points which represent the extremities of the other lines are as follows:

Point	Lat. Coord.	Dep. Coord.
C	+70	−20
D	+200	−100
E	−70	−40
F	−120	−200
G	−40	0
H	−100	+200

Coordinates are determined from the latitude and departure of individual courses. The latitude of a line is the projection of the line onto the Y axis and the departure of any line is the projection of that line onto the X axis. The value of these projections can be computed by using the length of the course or line and the angle which that line makes with the reference meridian. The latitude of a line is equal to the distance X the cosine of the bearing angle and the departure of the line is equal to the distance X the sine of the bearing angle.

Line	Bearing		Latitude	Departure
AB	N	E	+70	+170
CD	N	W	+130	−80
EF	S	W	−50	−160
GH	S	E	−60	+200

The coordinates of a point can therefore be obtained by adding the values of the latitude and departure of the line to the coordinate values at one extremity of the line.

The location of the origin or axis intersection will determine the coordinate values but it will never affect the value of the latitude and departure of a line. For example in Figure 4-2, assume the distance AB = 100.00′ and the bearing of line AB is N 45°00′E:

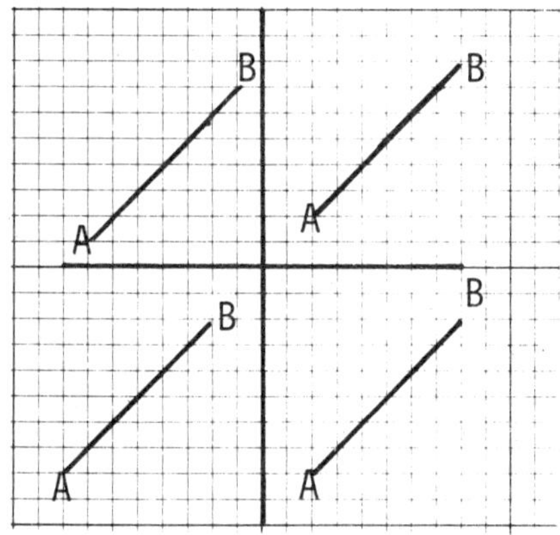

Figure 4-2. Relationship of latitude and departure to coordinates.

In all examples the latitude of AB = 100.00 $\times$ cosine 45° = +70.71′ and the departure of AB = 100.00′ $\times$ sine 45° = +70.71′. The coordinate values of B are, however, determined by the coordinates of A. If the line BA were being considered, obviously the bearing of BA would be S 45°00′W and the latitude and departure of line BA then have negative values.

The signs (±) are assigned to latitudes and departures according to their direction and not with respect to a certain or beginning point of the traverse. The latitude is N or (+) if the line has a northerly direction (NE or NW) and S or (-) for all lines having a southerly direction (SW or SE). Positive or E departure is applied to all lines having an easterly direction (NE or SE) and vice versa for all lines with a westerly direction (NW or SW). Latitudes and departures are computed for the purpose of checking for any error that might exist and for the computation of latitude coordinates and departure coordinates.

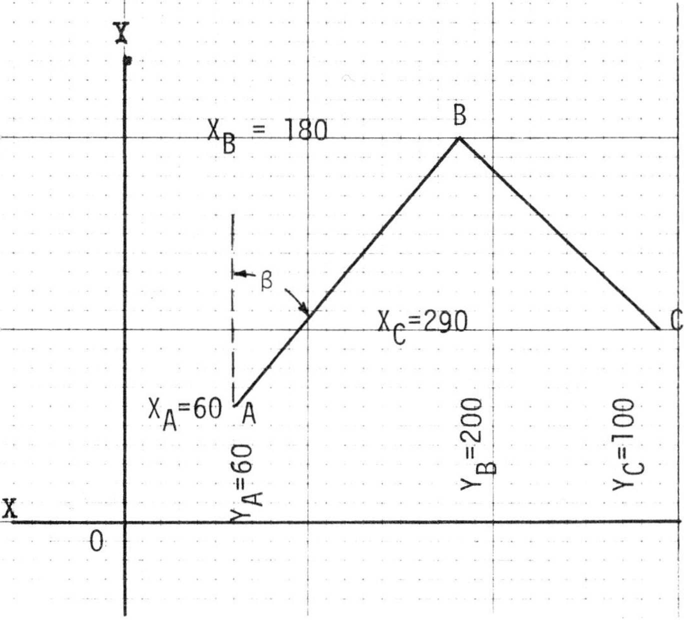

Figure 4-3. Computation of coordinates using latitude and departure.

Lat. Coord. B $= Y_B =$ Lat. Coord. A + latitude AB

$$= +60 +140 = +200$$

Dep. Coord. B $= X_B =$ Dep. Coord. A + departure AB

$$= +60 + 120 = +180$$

Latitude AB = Lat. Coord. B - Lat. Coord. A

$$= Y_B - Y_A = 200 - 60 = +140$$

Departure AB = Dep. Coord. B - Dep. Coord. A

$$= X_B - X_A = 180 - 60 = +120$$

Tangent β = tangent of bearing of line AB

$$= \frac{X_B - X_A}{Y_B - Y_A} = \frac{\text{departure AB}}{\text{latitude AB}}$$

$$\text{Distance AB} = \sqrt{(Y_B - Y_A)^2 + (X_B - X_A)^2}$$

$$= \frac{Y_B - Y_A}{\cos \beta} = \frac{X_B - X_A}{\sin \beta} = \frac{\text{Lat AB}}{\cos \beta} = \frac{\text{Dep AB}}{\sin \beta}$$

Considering the line BC, the latitude would equal $Y_C - Y_B = 100 - 200 = -100$. The departure of line BC would equal $X_C - X_B = 290 - 180 = +110$. The tangent of the bearing of line BC = +110/-100 = -1.10. Note the (-) sign would indicate a bearing of SE or NW but in this case the (-) sign indicates SE because the latitude of the line BC was negative. It is very important to keep track of the sign of the latitudes and departures. The $\tan^{-1} -1.10 = 47°43'34.7''$ or the bearing of line BC = S 47°43'34.7''E.

In a closed traverse the distance that is traversed north should equal the distance traversed south and the same principle applies to the east-west directions. Keep in mind that + and – signs are assigned to the latitude and departure of each course. Therefore, the summation (Σ) of the latitudes should be equal to zero and the summation (Σ) of the departures should also equal zero. Due to angular error and chaining errors, this very seldom occurs. By summing up the latitudes and departures, the error of closure may be determined before plotting any data and if the error is within the acceptable range, it can be distributed around the traverse.

The computations for the traverse may be done on a computer by one of several programs or it may be done with the aid of a desk or pocket calculator. The following form is suggested if a calculator is used.

Line - Bearing - Cos.Brg. - Sin.Brg. - Hor.Distance - Latitude - Departure
+   –   +   –

With this form, the sum of the + and – values are easily determined. The example shown on page 90 shows the calculation of a closed traverse which did not close.

Figure 4-4 shows the error that exists. A is the beginning point and A' is the end point. The linear error is the distance from A' to A and the bearing that from A' to A.

The direction of A' A can readily be determined by the signs of the Σ departures and the Σ latitudes. The linear error does not mean much by itself but when expressed as a ratio of the total horizontal distance around the traverse, it becomes the relative error of closure. If the value of the

Station Horizontal Distance	Bearing	Cos.	Sin.	Latitude +	Latitude −	Departure +	Departure −
A							
120.61'	N 86°00'W	0.069756	0.997564	8.4133			120.3162
B							
114.74'	N 56°50'W	0.547076	0.837083	62.7716			96.0469
C							
191.50'	N 42°00'E	0.743145	0.669131	142.3122		128.1385	
D							
156.81'	S 68°50'E	0.361082	0.932534		56.6213	146.2307	
E							
167.92'	S 21°00'W	0.933580	0.358368		156.7668		60.1771
A							

ΣH.D. = 751.58 $\quad\quad$ Σ=+213.4971 Σ=−213.3881 Σ=+274.3692 Σ=−276.5402

$$\Sigma\Sigma = +0.1090 \qquad\qquad \Sigma\Sigma = -2.1710$$

$$\text{Linear Error} = \sqrt{(0.1090)^2 + (-2.1710)^2}$$
$$= \sqrt{0.011881 + 4.713241}$$
$$= \sqrt{4.725122}$$
$$= 2.1737$$

$$\text{Relative error} = \frac{\text{Linear Error}}{\text{Hor. Distance}} = \frac{2.1737}{751.58} = \frac{(2.1737/2.1737)}{(751.58/2.1731)} = \frac{1}{345.76} = \frac{1}{350}$$

$$\text{Tangent } \beta = \frac{-2.1710}{0.1090} = 19.91743, \quad \tan^{-1} 19.91743 = 87°07'33''$$

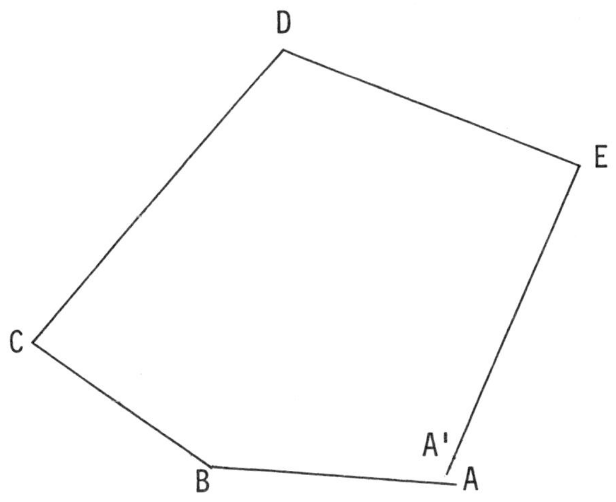

Figure 4-4. Closed traverse with linear error of closure.

linear error = 0.75 and the horizontal distance = 1250 feet, the relative error of closure then = 0.75/1250 = 1/1666. The relative error of closure is always expressed as a fraction with 1 in the numerator. This is done by dividing the numerator and the denominator by the numerator. The denominator is usually rounded off to the nearest 50 or 100 after the division is performed. Thus 1/1666 would be given as 1/1650 or 1/1700. In the sample traverse, the relative error of closure or precision is 1/350. This error may be proportioned around the traverse by one of several methods which will produce a mathematically closed figure. However, the surveyor should use his own judgement as to the size of the corrections and where they should be applied, based on field conditions. Under no circumstances should corrections be made for error that exists only because of gross human error.

The Compass Rule for adjusting a traverse assumes that the angle and distance measurements are made with equal precision. This rule is also the one most commonly used for adjusting traverses. The correction in latitude or departure to be applied to any one course is to the total error in latitude or departure as the length of that course is to the total distance around the traverse.

$$\frac{\text{Correction in latitude for AB}}{\text{closing error in latitude}} = \frac{\text{length of AB}}{\text{perimeter of traverse}}$$

$$\frac{\text{Correction in departure for AB}}{\text{closing error in departure}} = \frac{\text{length of AB}}{\text{perimeter of traverse}}$$

or

Correction in latitude for any course

$$= \frac{\Sigma \text{ latitudes}}{\text{perimeter of traverse}} \times \text{length of course}$$

Correction in departure for any course

$$= \frac{\Sigma \text{ departures}}{\text{perimeter of traverse}} \times \text{length of course}$$

Logically, the compass rule is not the best method of proportioning the error around a closed traverse. Assume for example that the first course of a traverse had a length of n feet and a bearing of N 00°00'E. If there is an error in the departures in closing the traverse, a departure correction is given to this course which had no departure and a new bearing is then computed even though the bearing of this first course was assumed to be correct. It is usually assumed that distance and angular measurements are of equal precision, but it is more likely that errors occur in chaining and that there is very little error in angular measurements which are checked by summing the interior or deflection angles of a closed traverse.

The traverse data previously computed for Figure 4-4 corrected by the compass rule would be as follows.

Σ H.D. = 751.58, Σ Lat. = 0.1090, Σ Dep. = -2.1710

Correction in Lat. = (0.1090/751.58) $\times$ length of course
= 0.000145 $\times$ length of course

Correction in Dep. = (2.1710/751.58) $\times$ length of course
= 0.002889 $\times$ length of course

Course	H.D.	Correction in Lat.	Correction in Dep.
AB	120.61	120.61 $\times$ 0.000145 = 0.0175	120.61 $\times$ 0.002889 = 0.3484
BC	114.74	114.74 $\times$ 0.000145 = 0.0166	114.74 $\times$ 0.002889 = 0.3314
CD	191.50	191.50 $\times$ 0.000145 = 0.0278	191.50 $\times$ 0.002889 = 0.5532
DE	156.81	156.81 $\times$ 0.000145 = 0.0228	156.81 $\times$ 0.002889 = 0.4530
EA	167.92	167.92 $\times$ 0.000145 = 0.0243	167.92 $\times$ 0.002889 = 0.4850
		Σ = 0.1090	Σ = 2.1710

It should be noted that no sign value has been given to the corrections but that the sum of the corrections is equal to the original error of both the latitudes and departures. If the sum of the corrections does not equal the total error of each, the computations must be checked again for mathematical errors.

The Transit Rule assumes that the angular measurements are of a higher precision than are the distance measurements. The correction applied to latitude or departure of a course using the transit rule is to the total error in latitudes or departures as the latitude or departure of the course is to the sum of all latitudes or departures without regard to the algebraic signs.

$$\frac{\text{Correction in latitude for AB}}{\text{closing error in latitude}} = \frac{\text{latitude of AB}}{\text{arithmetical sum of all latitudes}}$$

$$\frac{\text{Correction in departure for AB}}{\text{closing error in departure}} = \frac{\text{departure of AB}}{\text{arithmetical sum of all departures}}$$

$$\frac{\text{Correction in latitude for any course}}{} = \frac{\Sigma \text{ latitudes} \times \text{latitude of course}}{\text{arithmetical sum of all latitudes}}$$

$$\frac{\text{Correction in departure for any course}}{} = \frac{\Sigma \text{ departures} \times \text{departure of course}}{\text{arithmetical sum of all departures}}$$

The corrections of the traverse computed for Figure 4-4 corrected by the transit rule would be as follows.

Σ latitudes $= 0.1090$ arithmetic sum of latitudes $= (+)213.4971 + (-)213.3881 = 426.8852$

Σ departures $= -2.1710$ arithmetic sum of departures $= (+)274.3692 + (-)276.5402 = 550.9094$

Correction in latitude $= \dfrac{0.1090}{426.8852} \times$ latitude of course $= 0.000255 \times$ latitude of course

Correction in departure $= \dfrac{-2.1710}{550.9094} \times$ departure of course $= 0.003941 \times$ departure of course

Course	Latitude	Departure	Correction in latitude	Correction in departures
AB	+8.4133	-120.3162	$8.4133 \times 0.000255 = 0.00215$	$120.3162 \times 0.003941 = 0.4741$
BC	+62.7716	-96.0469	$62.7716 \times 0.000255 = 0.01603$	$96.3162 \times 0.003941 = 0.3785$
CD	+142.3122	+128.1385	$142.3122 \times 0.000255 = 0.03634$	$128.1385 \times 0.003941 = 0.5050$
DE	-56.6213	+146.2307	$56.6213 \times 0.000255 = 0.0145$	$146.2307 \times 0.003941 = 0.5763$
EA	-156.7668	-60.1771	$156.7668 \times 0.000255 = 0.04003$	$60.1771 \times 0.003941 = 0.2371$
			$\Sigma = 0.1090$	$\Sigma = 2.1710$

The corrections in latitudes were carried out to 5 places in order to have the sum of all of the corrections equal to the total error in latitude. It should be obvious that these corrections, which would normally be rounded off to 2 or 3 places when computing the corrections, are small because of the very small error in the latitudes.

A third method of adjusting a traverse is by the Crandall method, the results of which are similar to those obtained using the compass rule. The angular error is distributed first in equal amounts to all angles and then the linear error is distributed to all measurements by a weighted least square procedure which can only be done with a programable calculator. The corrections in latitude and departure for the traverse shown in Figure 4-4 are shown only for the sake of comparison. A fourth method of balancing a traverse is by the method of least-squares, a complicated procedure beyond the scope of this text.

When all of the corrections have been computed, they are applied to the original computed values. Add the corrections to the latitudes or departures whose sums are the smallest and subtract them from the latitudes and departures whose sums are the largest. A better method to use is as follows.

Corrected latitude of a course

$$= \text{computed latitude} - \left[\left(\begin{matrix} \text{sign of} \\ \text{error} \end{matrix} \right) \begin{matrix} \text{Computed correction} \\ \text{of latitude} \end{matrix} \right]$$

Corrected departure of a course

$$= \text{computed departure} - \left[\left(\begin{matrix} \text{sign of} \\ \text{error} \end{matrix} \right) \begin{matrix} \text{Computed correction} \\ \text{of departure} \end{matrix} \right]$$

In the problem being used as an example, the sign of the error used in computing the corrected latitudes would be + because the Σ latitudes = +0.1090 and the sign of the error used to compute the corrected departures would be – because the Σ departures = –2.1710.

Figure 4-5 gives a comparison of the corrected latitudes and departures values computed by the Compass Rule, Transit Rule and Crandall's method.

In some instances it may be necessary to compute new adjusted distances and adjusted bearings using the corrected latitudes and departures. Figure 4-6 shows these adjusted values using the traverse data which has been balanced by the Compass Rule.

Adjusted Bearings and Distances

Course	Adjusted Latitude	Adjusted Departure	Adjusted Bearing	Adjusted Distance
AB	+8.2958	–119.9678	N 85°59'48"W	120.26
BC	+62,7550	–95,7155	N 56°44'58"W	114.45
CD	+142.2844	+128.6917	N 42°07'42"E	191.85
DE	–56.6441	+146.6837	S 68°53'07"E	157.24
EA	–156.7911	–59.6921	S 20°50'33"W	167.77

The above adjusted bearings and distances were computed using the following relationships.

$$\frac{\text{Adjusted departure}}{\text{Adjusted latitude}} = \text{Tangent of adjusted bearing}$$

$$\text{Adjusted Distance} = \frac{\text{Adjusted latitude}}{\text{Cos Adjusted Bearing}} = \frac{\text{Adjusted departure}}{\text{Sin Adjusted Bearing}}$$

$$\text{Adjusted Distance} = \sqrt{(\text{Adjusted latitude})^2 + (\text{Adjusted departure})^2}$$

Course	Unadjusted		Latitude Adjustment	Departure Adjustment	Method of Adjustment
	Latitude	Departure			
AB	+8.4133	−120.3162	+8.4133 − (+)0.0175 = +8.3958 +8.4133 − (+)0.0022 = +8.4111 +8.3710	−120,3162 − (−)0.3484 = −119.9678 −120.3162 − (−)0.4741 = −119.8421 −119.7116	Compass Transit Crandall's
BC	+62.7716	−96.0469	+62.7716 − (+)0.0166 = +62.7550 +62.7716 − (+)0.0160 = +62.7556	−96.0469 − (−)0.3314 = −95.7155 −96.0469 − (−)0.3785 = −95.6684 −95.5967	Compass Transit Crandall's
CD	+142.3122	+128.1385	+142.3122 − (+)0.0278 = +142.2844 +142.3122 − (+)0.0363 = +142.2759 +142.6836	+128.1485 − (−)0.5532 = +128.6917 +128.1385 − (−)0.5050 = +128.6435 +128.4729	Compass Transit Crandall's
DE	−56.6213	+146.2307	−56.6213 − (+)0.0228 = −56.6441 −56.6213 − (+)0.0145 = −56.6358 −56.9034	+146.2307 − (−)0.4530 = +146.6837 +146.2307 − (−)0.5763 = +146.8070 +146.9593	Compass Transit Crandall's
EA	−156.7668	−60.1771	−156.7668 − (+)0.0243 = −156.7911 −156.7668 − (+)0.0400 = −156.8068 −156.6286	−60.1771 − (−)0.4850 = −59.6921 −60.1771 − (−)0.2371 = −59.9400 −60.1240	Compass Transit Crandall's
			Σ−Lat.=213.4352 Σ+Lat.=213.4352 Σ−Lat.=213.4426 Σ+Lat.=213.4426 Σ−Lat.=213.5320 Σ+Lat.=213.5320	Σ−Dep.=275.3754 Σ+Dep.=275.3754 Σ−Dep.=275.4505 Σ+Dep.=275.4505 Σ−Dep.=275.4322 Σ+Dep.=275.4323	Compass Transit Crandall's

Figure 4-5. Latitudes and departures adjusted by Compass, Transit and Crandall's Rule.

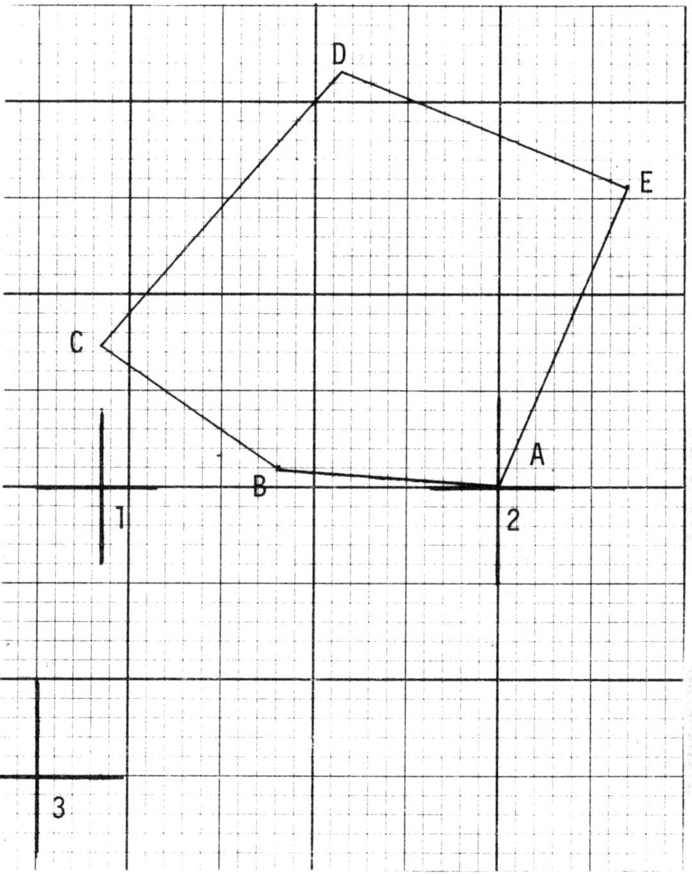

Figure 4-6. Position of traverse with respect to origin.

The origin or the intersection of the meridian and parallel (Y and X axis) should be chosen, if possible, so that the entire traverse will lie in the first quadrant and then all of the coordinate values will be positive. One method is to pass the reference meridian through the most westernly point of the traverse and the reference parallel through the most southerly point. This will give these points a departure coordinate and latitude coordinate of zero respectively. Sometimes this is not convenient and it is more desirable to compute the coordinates of all stations in the traverse in relation to the beginning point. The coordinates of the beginning point may be assumed to be 0 and 0 or the coordinate values of the beginning point may be assigned values large enough to place the complete traverse in the first quadrant, making all values positive. Since coordinates are the summation of latitudes and departures, the coordinate of any point is equal to the coordinate (latitude or departure) of the first point plus the latitude or departure of every course to the point itself. Be sure to use the proper sign of the latitude or the departure of each course being considered.   In other words, the latitude coordinate of E = the latitude coordinate of A ± latitude AB ± latitude BC ± latitude CD ± latitude of DE.

Examining the data given for Figure 4-4 on page 90 will determine that station C is the most westerly point of the traverse and A is the most southerly point of the traverse. Assuming that the departure coordinate of C is 0 and the latitude coordinate of A is 0 will place the figure in the first quadrant with the origin at (1) as shown in Figure 4-6. If the coordinates of A are assumed to be 0 and 0, the origin is then located at (2) but this would result in some points having negative coordinate values. The origin could also be placed at (3) by assuming the latitude coordinate of A = 150 and the departure coordinate of A = 250.

The table below shows the coordinates of each station assuming the origin to be at the three points shown in

Station	Latitude	Departure	Origin at (1)		Origin at (2)		Origin at (3)	
			L.C.	D.C.	L.C.	D.C.	L.C.	D.C.
A			0	+215.6833	0	0	+150.000	250.000
	+8.3958	−119.9678						
B			+8.3958	+95.7155	+8.3958	−119.9678	+158.3958	+130.0322
	+62.7550	−95.7155						
C			+71.1508	0	+71.1508	−215.6833	+221.1508	+34.3164
	+142.2844	+128.6917						
D			+213.4352	+128.6917	+213.4352	−86.9916	+363.4352	+163.0084
	−56.6441	+146.6837						
E			+156.7911	+275.7011	+156.7911	+59.6921	+306.7911	+309.6921
	−156.7911	−59.6921						
A			0	+215.6833	0	0	+150.000	+250.0000

Figure 4-6. The adjusted latitudes and departures were computed using the Compass Rule.

Since plotting of traverse data is most commonly done using coordinates, the advantages should be emphasized. The shape and size can be determined before plotting so that the appropriate scale may be selected. For a closed traverse, the error of closure can be computed and the field work checked or the traverse balanced before the plotting is begun. The most important fact is that the accuracy of plotting a point does not depend on the accuracy of the plotting of any of the preceding points.

If for any reason it is impossible or impractical to determine by field measurements the lengths or angles necessary to compute the bearing of any course in a closed traverse, the missing data may be computed if no more than two quantities are missing. Typical problems are as follows:

1. Missing length and bearing of one side.

2. Missing length of one side and bearing of second side.

3. Missing lengths of two sides.

4. Missing bearings of two sides.

It must be emphasized that these types of solutions should be resorted to only if there is no other way of obtaining the missing data. There is no means of determining the error of closure and if there are any errors in the field data, all errors are thrown into either the calculated lengths or bearings of the lines. A hint to solving such problems is to isolate the parts that are unknown from the rest of the traverse. This means that the sides of a closed traverse may have to be shifted because to isolate the parts with unknown values they must be adjacent to each other.

1. Length and bearing of one side unknown.
   Given: Lengths and bearing of AB, BC, CD and DE.
   Find: Length and bearing of EA.

Figure 4-7 illustrates the traverse with the dashed line EA representing the unknown part of the traverse.

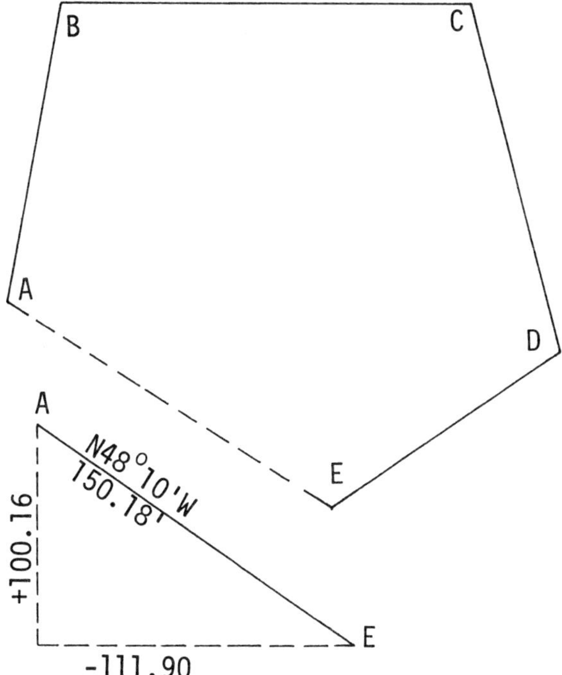

Figure 4-7. Traverse with length and bearing of one side missing.

Solution—Solve for the latitudes and departures of AB, BC, CD and DE then solve for the summation of the latitudes and departures. Then the distance $EA = \sqrt{\Sigma(\text{lat})^2 + \Sigma(\text{dep})^2}$ and the tan of the bearing angle of EA = Σ dep/Σ lat. The following data is for the traverse shown in Figure 4.7.

Course	Distance	Bearing	(+) Latitudes (−)		(+) Departures (−)	
AB	100.00′	N 16°20′E	95.9642		28.1225	
BC	140.00′	S 84°43′E		12.8913	139.4052	
CD	120.00′	S 12°30′E		117.1555	25.9727	
DE	105.00′	S 51°00′W		66.0786		81.6003
EA	?	?				

$$\Sigma = +95.9642 \quad \Sigma = -196.1255 \qquad \Sigma = +193.5005 \quad \Sigma = -81.6003$$
$$\Sigma\Sigma = -100.1613 \qquad\qquad \Sigma\Sigma = +111.9002$$

$$\text{Tan Brg} \angle = \frac{+111.9002}{-100.1613} = -1.117199$$

$$\angle = 48.16842°$$

The Σ latitudes = −100.1613, but in a closed traverse the Σ latitudes should = 0. Hence, the latitude of EA = +100.1613 and the departure of EA must = −111.9002.

$$\text{Distance} = \frac{100.1613}{\cos 48.16842°} = \frac{111.9002}{\sin 48.16842°} = \underline{150.1796}$$

Bearing EA = $\underline{\text{N } 48°10′06''\text{W}}$

2. Length of one side and bearing of another side unknown.

    Given: Lengths and bearing of AB, BC, and CD, the bearing of DE and length of EA.

    Find: Length of DE and bearing of EA.

Figure 4-8 illustrates the isolation of the unknown parts from the traverse with the line AD.

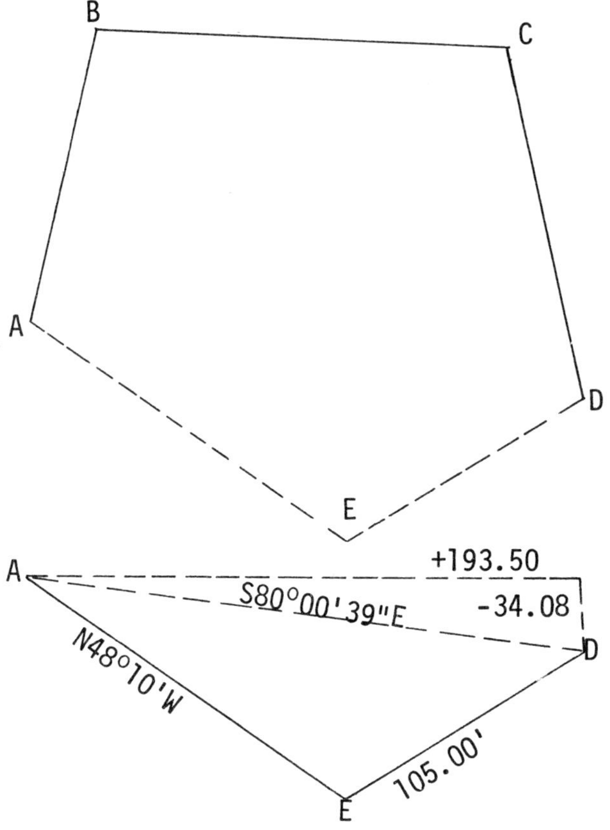

Figure 4-8. Traverse with length of one side and bearing of another side missing.

Course	Distance	Bearing	(+) Latitudes (−)		(+) Departures (−)	
AB	100.00′	N 16°20′E	95.9642		28.1125	
BC	140.00′	S 84°43′E		12.8913	139.4052	
CD	120.00′	S 12°30′E		117.1555	25.9728	
DE	105.00′	?				
EA	?	N 48°10′06″W				

$$\Sigma=+95.9642 \quad \Sigma=-130.0468 \quad \Sigma=+193.5005$$
$$\Sigma\Sigma = -34.0826 \qquad\qquad \Sigma\Sigma = +193.5005$$

−34.0827 represents the latitude of line AD and +193.5005 represents the departure of line AD, which isolates the unknown parts DE and EA from the rest of the traverse.

$$\text{Tan of Brg of AD} = \frac{+193.5005}{-34.0827} = -5.677386$$

Bearing of AD = S 80.010535° E

$$\text{Distance AD} = \frac{193.5005}{\sin 80.010535°} = 196.479 \text{ feet}$$

Thus in the triangle ADE, two sides and an interior angle opposite one of the sides are known and the law of sines is applied. Note that angles have been changed to decimals to facilitate the use of a calculator.

$\angle$ EAD 80°00′38″ − 48°10′ = 31.8422°

$$\frac{105.00}{\sin 31.8422°} = \frac{196.47915}{\sin \angle DEA} = \frac{AE}{\sin \angle ADE}$$

$$\sin \angle DEA = \frac{196.47915}{105.00} \times \sin 31.843893° = 0.98722587$$

$\angle$ DEA = 80.83167° or 99.16783°.

Inspection would show that $\angle$ DEA = 99.16325° and

$\angle$ ADE = 180° − (31.84220° + 99.16325°) = 48.98996°

Distance EA = 150.191 feet

Bearing DE = $180° - (80.01105° + 48.99463°) = 50.9995°$
$$= S\ 50°58'58''W$$

3. Length of two sides are unknown.
   Given: Lengths and bearings of AB, BC, EA and bearings of CD and DE.
   Find: Lengths of CD and DE.

Figure 4-9 illustrated line CE cutting off the two unknown parts of the closed traverse.

Course	Distance	Bearing	(+) Latitudes (−)		(+) Departures (−)	
AB	100.00′	N 16°20′E	95.9642		28.1225	
BC	140.00′	S 84°34′E		12.8913	139.4052	
CD	?	S 12°30′E				
DE	?	S 51°00′W				
EA	150.18′	N 48°10′06″W	100.1613			111.9001

$$\Sigma = +196.1255 \quad \Sigma = -12.8113 \qquad \Sigma = +167.5277 \quad \Sigma = -111.9001$$
$$\Sigma\Sigma = +183.2342 \qquad\qquad \Sigma\Sigma = +55.6276$$

Tan of bearing of CE $= \dfrac{+55.6276}{+183.2342} = 0.3035874$

Bearing of CE = S 16.8876306°W

Distance CE $= \dfrac{55.6276}{\text{sine } 16.8876°} = 191.4920$ feet

∢ DCE 29.3876°

∢ CED 51°00′ − 16.8876° = 34.11237°

∢ CDE = $180° - (29.3876° + 34.1124°) = 116.500°$

In triangle CDE, two angles and a side opposite one of the interior angles is known. Hence, the two sides that are

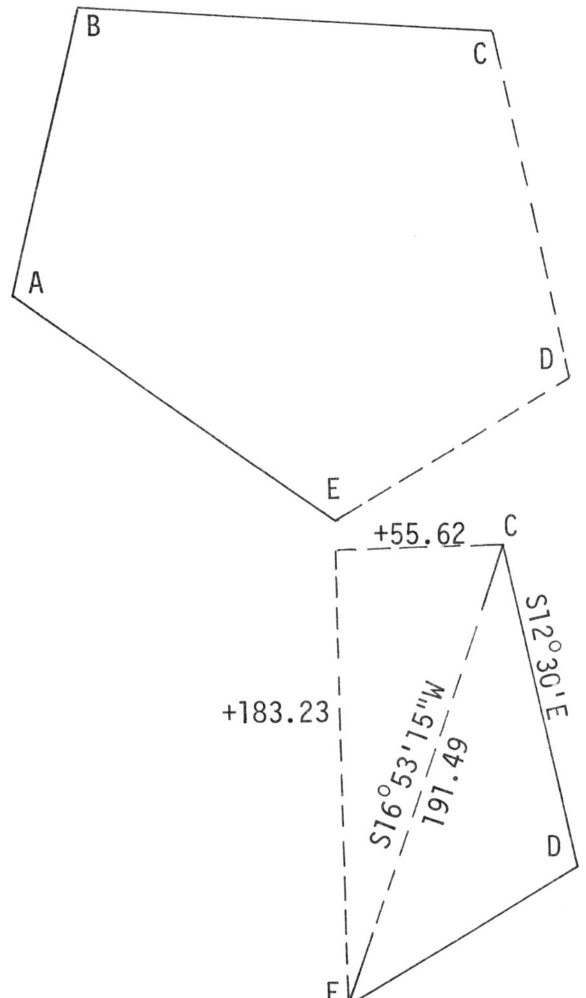

Figure 4-9. Traverse with length of two sides missing.

unknown can be solved for using the law of sines.

$$\frac{191.4920}{\sin 116.500°} = \frac{CD}{\sin 34.1124°} = \frac{DE}{\sin 29.3876°}$$

CD = 120.00 feet   DE = 104.99 feet

4. Bearing of two sides unknown.
   Given: Lengths and bearing of AB, DE, EA and lengths of BC and CD.
   Find: Bearing BC and bearing CD.

Figure 4-10 illustrates the solution of this type of problem.

Course	Distance	Bearing	(+) Latitudes (−)		(+) Departures (−)	
AB	100.00′	N 16°20′E	95.9642		28.1225	
BC	140.00′	?				
CD	120.00′	?				
DE	105.00′	S 51°00′W		66.0786		81.6003
EA	150.18′	N 48°10′06″W	100.1613			111.9001

Σ=+196.1255 Σ=−66.0786 Σ=+28.1225 Σ=−193.5004

$\Sigma\Sigma$ = +130.0469 $\Sigma\Sigma$ = −165.3779

Tan of bearing BD $= \dfrac{-165.3779}{+130.0469} = -1.271679$

Bearing BD = S 51.8198°E

Distance BD $= \dfrac{165.3779}{\sin 51.8198} = 210.3855$ feet

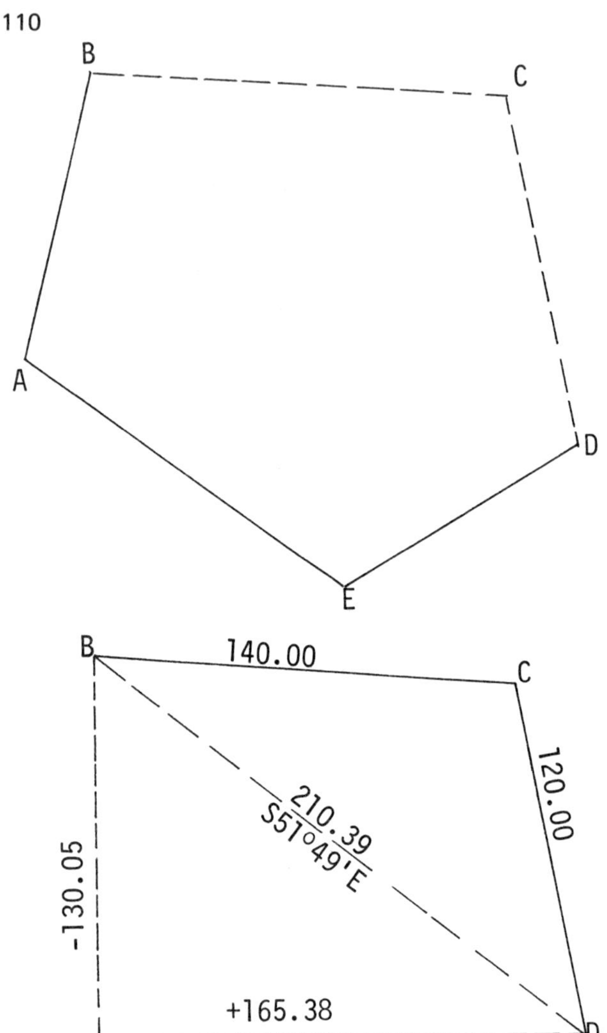

Figure 4-10. Traverse with bearings of two adjacent sides missing.

In triangle BCD, three sides are known and any one of the three interior angles can be solved where

$$S = \text{½ perimeter} = \frac{a + b + c}{2}.$$ S in this problem is equal to

$$\frac{140.00 + 120.00 + 210.3855}{2} = 235.1927$$

$$\text{Versine} \angle C = \frac{2(S-b)(S-c)}{bd}$$

$$= \frac{2(235.1927-140.00)(235.1927-120.00)}{(140.00)(120.00)}$$

$$= 1.305417$$

Versine $\angle C$ = 1 – cosine $\angle C$:

$$\text{cosine} \angle C = 1 - 1.305417 = -0.30547:$$

$$\angle C = 107.7833°$$

$$\text{Versine} \angle B = \frac{2(235.1927-140)(235.1927-210.3855)}{(210.3855)(140.00)}$$

$$= 0.160349$$

Versine $\angle B$ = 1 – cosine $\angle B$:

$$\text{cosine} \angle B = 1 - \text{versine} \angle B = 0.839651$$

$$\angle B = 32.89677°$$

$$\angle D = 39.3199°$$

$$\Sigma = 180.00°$$

Bearing DC = 51.8196° – 39.3199° = 12.4997°
Bearing DC = N 12°30′W
Bearing BC = 51.8198° + 32.8968° = 84.7166°
Bearing BC = S 84°43′E

Coordinates are very useful in solving for the point of intersection of two lines, a line and a curve or two curves if the equation for each line and or curve is known. Equations of straight lines may be expressed in terms of the slope. The slope is determined easily if the coordinates of two points, preferably at the extremities of the line are known. The slope of a straight line is given as $m = \tan \alpha$ when α is measured from the positive X axis. The angle α is the complement of the angle β which is the bearing of the line. Hence the slope of the line $= m = \cot \beta$. The equation of a straight line, when the coordinates of two points (X_1,Y_1) and (X_2,Y_2) are known is $\dfrac{Y-Y_1}{Y_2-Y_1} = \dfrac{X-X_1}{X_2-X_1}$. X and Y represent the coordinates of a point which are not known but which lies on this line. Figure 4-11 shows the relationship between the angles α and β.

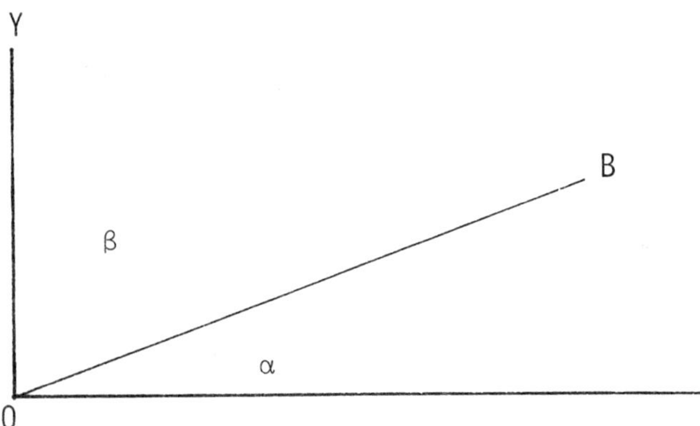

Figure 4-11. Relationship of slope of line to bearing of line.

The equation of a straight line then becomes $Y - Y_1 = \cot \beta(X-X_1)$ when the bearing angle and the coordinates of one point (X_1, Y_1) are known.

$$(1) \quad \frac{Y-Y_1}{Y_2-Y_1} = \frac{X-X_1}{X_2-X_1}$$

$$(2) \quad Y - Y_1 = \frac{Y_2-Y_1}{X_2-X_1}(X-X_1)$$

$$(3) \quad Y - Y_1 = \frac{\text{latitude}}{\text{departure}}(X-X_1)$$

$$(4) \quad Y - Y_1 = \cot \beta(X-X_1)$$

If the bearing is not known (it is not necessary to know the bearing) and the coordinates of two points are known it is more convenient to express the line in the form of equation (1). This equation uses (X,Y) as the unknown coordinates of a point on that line. If this line is intersected by a second line, the equation of the second line will have a point of unknown coordinates (X,Y) and since (X,Y) represent the same point, this point is the point of intersection. Two equations with two unknown values are most easily solved simultaneously. Figure 4-12 illustrates the data given below.

Given:	Point	Lat. Coordinate	Dep. Coordinate
	A	50 (Y_1)	30 (X_1)
	B	20 (Y_2)	100 (X_2)
	C	40 (Y_1)	80 (X_1)
	D	10 (Y_2)	10 (X_2)

114

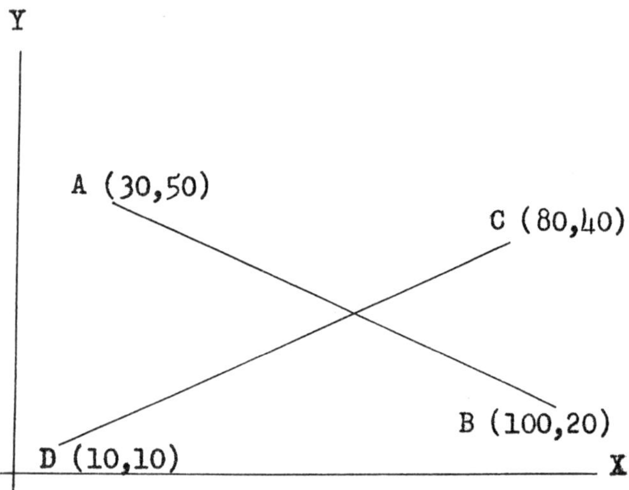

Figure 4-12. Coordinates of point of intersection of two
lines.

Find: The coordinates of the point of intersection of
lines AB and line CD. Using the equation $\dfrac{Y-Y_1}{Y_2-Y_1} = \dfrac{X-X_1}{X_2-X_1}$,

two equations one each for line AB and CD are obtained.

Line AB:

$$\frac{Y-50}{20-50} = \frac{X-30}{100-30}$$

$$70Y - 3500 = -30X + 900$$

(1) $70Y + 30X = 4400$

Line CD:

$$\frac{Y-40}{10-40} = \frac{X-80}{10-80}$$

$$-70Y + 2800 = -30X + 2400$$

(2) $70Y - 30X = 400$

Adding the two equations (1) and (2) together gives 140Y = 4800. Solving for Y which is the latitude coordinate of the point of intersection: Y = 34.285. Substituting this value of Y in equation (1) gives:

$$(70)(34.285) + 30X = 440, \quad 30X = 2000.00 \text{ or } X = 66.666$$

This type of solution is easily applied to the problem of finding the point of intersection of the ℄ (centerline) of a road and a property line, and in finding the center of a section, which is the point of intersection of lines connecting opposite quarter corners.

Simultaneous equations need not be restricted to the solution of finding the point of intersection of lines but can also be used in traverses if two equations can be written which involve only two unknowns. Take the problem of a traverse which has two sides of unknown length but the sides are not adjacent to each other. The unknown lengths may be found by any one of three different methods. Such a traverse is shown in Figure 4-13.

Course	Distance	Bearing	(+) Latitudes	(−)	(+) Departures	(−)
AB	690.00′	N 09°30′W	680.5371			113.8828
BC	?	N 56°50′W				
CD	675.00′	S 56°14′W		375.1732		561.1329
DE	?	S 02°03′E				
EA	1083.00′	S 89°35′E		7.8757	1082.9714	

$$\Sigma = +680.5371 \quad \Sigma = -383.0489 \quad \Sigma = +1082.9714 \quad \Sigma = -675.0157$$

$$\Sigma\Sigma = +297.4882 \qquad\qquad \Sigma\Sigma = +407.9557$$

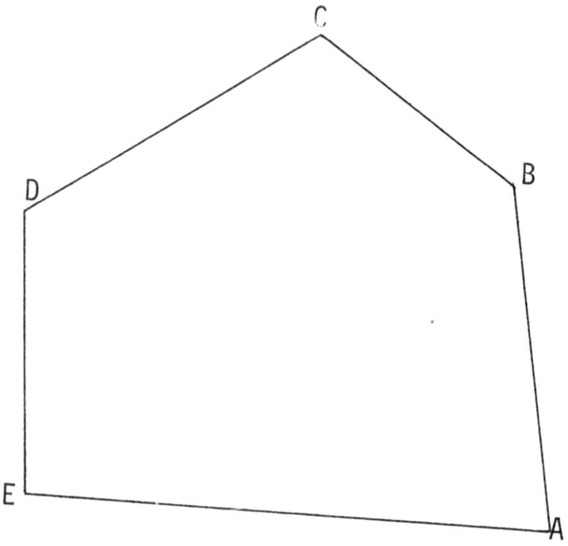

Figure 4-13. Traverse with two sides which are not adjacent having lengths unknown.

Solution I

Previously described problems of this type have been solved by isolating the unknown parts of the traverse, but this is not directly possible in this case because the unknown sides are not adjacent. However, DE may be shifted without changing its length or bearing to a position following line BC. Care must be taken in moving a side to maintain a closed figure. Side CD is also shifted to a new position as shown in Figure 4-14.

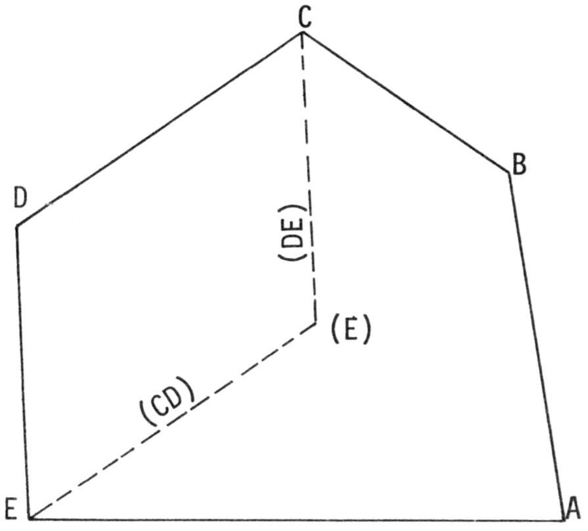

Figure 4-14. Shifting of two unknown sides to make them adjacent. Solution 1.

The summation of the latitudes and departures now represents the latitude and the departure of line (E)B which isolates the two unknown quantities.

$$\text{Tangent of bearing (E)B} = \frac{\text{departure}}{\text{latitude}} \quad \frac{407.9557}{297.4882} = 1.371333$$

$$\text{Tangent } \angle = 1.371333$$

$$\angle = 53.899833^\circ \qquad \text{Bearing (E)B} = \text{N } 53.8998^\circ \text{ E}$$

$$\text{Distance (E)B} = \frac{407.9557}{\sin 53.8998^\circ} = 504.9030 \text{ feet}$$

Since all bearings are now known for the three sides of the triangle C(E)B, the interior angles are computed as follows:

$$\angle\ C(E)B = 55.9498°, \quad \angle\ (E)BC = 69.2668°,$$

$$\angle\ BC(E) = 54.7833°$$

Side (E)B was computed using the latitudes and departures. The law of sines can be applied to solve for the unknown distances.

$$\frac{504.9030}{\sin 54.7833°} = \frac{BC}{\sin 55.9498°} = \frac{DE}{\sin 69.2668°}$$

$$BC = 512.05 \text{ feet} \qquad DE = 577.99 \text{ feet}$$

Solution II

The solution for the unknown distances can be made without shifting the sides but merely by extending side BC and side ED to the point 0 where they intersect as shown in Figure 4-15.   The unknown distance BC = 0B − 0C and DE = E0 −0D. The solution is then one of solving for two sides of a triangle when all interior angles and one side are known.

$$\text{Tangent of bearing EB} = \frac{\text{departure EB}}{\text{latitude EB}} = \frac{\text{dep.EA} + \text{dep.AB}}{\text{lat.AB} + \text{lat.EA}}$$

$$= \frac{1082.9714 + (-113.8828)}{680.5371 + (-7.8757)}$$

$$= \frac{969.0886}{672.6614} = 1.440678$$

Bearing EB = N 55.2348°E

$$\text{Distance EB} = \frac{\text{latitude EB}}{\cos 55.2348} = 1179.6635 \text{ feet}$$

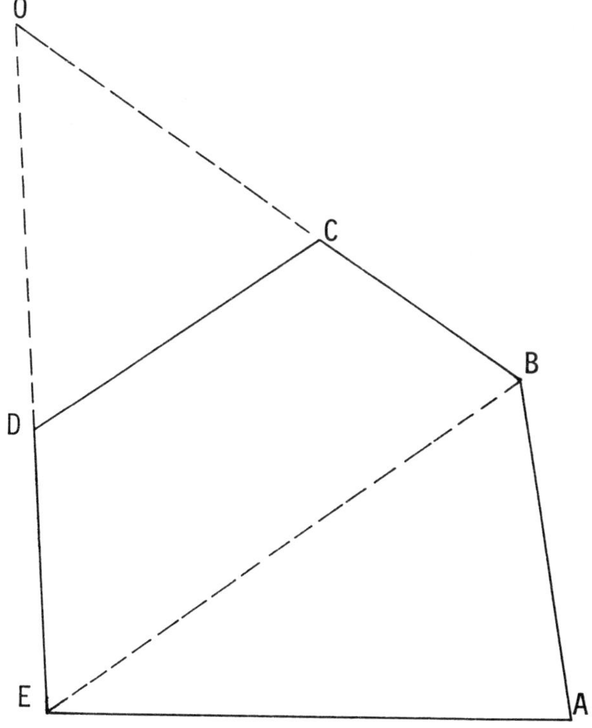

Figure 4-15. Solution of traverse without shifting sides. Solution 2 and 3.

Triangle EOB

EB = 1179.6635 feet

$\angle$ BEO = 57.2841°

$\angle$ BOE = 54.7833°

$\angle$ EBO = 67.9318°

$$\frac{1179.6635}{\sin 54.7833} = \frac{EO}{\sin 67.9318} = \frac{BO}{\sin 57.2848}$$

EO = 1338.1807 feet   BO = 1214.8820 feet

BC = OB − OC

BC = 1214.7925 − 702.8278

BC = **512.054** feet

Triangle DCO

DC = 675.00 feet

$\angle$ CDO = 58.2833°

$\angle$ DCO = 66.9333°

$\angle$ COD = 54.7833°

$$\frac{675.00}{\sin 54.7833} = \frac{CO}{\sin 58.2833} = \frac{DO}{\sin 66.9333}$$

CO = 702.8278 feet   DO = 760.1601 feet

DE = EO − OD

DE = **1338.1507** − 760.1601

DE = **577.9906** feet

Solution III

This solution involves the use of simultaneous equations written in the form of a trigonometric function of an angle. When the length and bearing of a line are known, the latitude and departure are expressed as the (cos ∡)(distance) and (sin ∡)(distance) respectively. In this example, if the distance is unknown, the latitude and departure are expressed as the numerical value of the cosine and sine of the angle X the distance written as BC or DE.

Course	Distance	Bearing	(+) Latitudes (−)		(+) Departures (−)	
AB	690.00′	N 09°30′W	680.5371			113.8828
BC	? (BC)	N 56°50′W	0.54707 BC*			0.83708 BC**
CD	675.00′	S 56°14′W		375.1732		561.1329
DE	? (DE)	S 02°03′E		0.99936 DE	0.035772 DE	
EA	1083.00′	S 89°35′E		7.8757	1082.9714	

* 0.54707 = cosine 56°50′
** 0.83708 = sine 56°50′

The sum of the latitudes contains two unknowns BC and DE. These same unknowns occur in the summation of the departures. The summation of the latitudes and departures equal zero in a closed traverse. Since these two summations contain two knowns, the value of BC and DE may be obtained by solving the equations simultaneously with particular care given to the signs of the trigonometric functions.

$$\Sigma \text{ latitudes} = 680.5371 + 0.54707BC - 375.1732$$
$$- 0.9936DE - 7.8957 = 0$$
$$= 297.4882 + 0.54707BC - 0.99936DE = 0$$

$$\Sigma \text{ departures} = 0.035772DE + 1082.9714 - 113.8828$$
$$- 0.83707BC - 561.1329 = 0$$
$$= 407.9557 + 0.035772DE - 0.83708BC = 0$$

Rearranging the equations gives the following:

(1) Σ Latitudes $0.54707BC - 0.99936DE = -297.4882$

(2) Σ Departures $-0.83707BC + 0.035772DE = -407.9557$

(3) Dividing (1) by 0.99936 gives:

   $0.54742BC - DE = -297.6787$

(4) Dividing (2) by 0.035772 gives:

   $-23.400145BC + DE = -11404.3302$

Adding (3) and (4):

   $-22.852725\,BC = -11702.0089$

   $BC = 512.06$ feet

Substituting in (2)

   $DE = 577.99$ feet

1.   At point A on a ridge whose elevation is 4000 feet, you drill a hole 2 inches in diameter to a depth of 210 feet (point D). At this elevation (D) you strike a rich vein of ore which you wish to remove through a tunnel beginning at point C which is the tunnel entrance. Point C is reached by the following traverse from Point A.

Course	Slope distance	Vertical angle	Bearing
AB	360.00 feet	-50%	S 25°00′W
BC	390.00 feet	-30%	N 85°00′W

Find the bearing of CD and the slope of CD in percent.

2.   Lines AC and BD are long enough to intersect. Bearing AC = N 41°30′E and the bearing of BD = N 83° 20′W. Find the latitude and departure coordinates of the point of intersection of line AC and line BD.

Station	Departure Coordinate	Latitude Coordinate
A	40	40
B	100	60

3.   Find the length and bearing of a line from the mid-point of line CD to the midpoint of line AB.

Station	Departure Coordinate	Latitude Coordinate
A	80	10
B	30	115
C	140	20
D	100	160

4.   You wish to relocate a 1/16 corner which will be called B and which is known to be 1328 feet due east of the section corner which is point A. Starting at point A, you run a line N 90°00′E a distance of 800 feet to point C where you are unable to continue on this bearing because of an obstruction. At C you turn a left deflection angle of 8°25′ and continue on 600 feet to point D.   Find the distance you would measure from D and the deflection angle you would turn at D to set B (1/16 corner).

5.   Write the two equations which when solved will give the latitude and departure coordinates of B using the data given below.

Latitude coordinate of A = 1000
Departure coordinate of A = 3300
Slope distance from A to B = 196.43 feet
Zenith angle sighting from A to B = 278°21'42"
Azimuth of line AB = 202°04'36"

6.   Find the Latitude and Departure Coordinates of the center of the circumscribed circle of Triangle ABC. (Hint   Center of circumscribed circle is located by the inersection of the perpendicular bisectors of the sides of the triangle).

Triangle ABC
Side AB = 160.00 feet
Side BC = 200.00 feet
Angle CAB = 55°
Angle ACB = 40°
Bearing AB = N 30°00'E

7. Find the slope distance from station B to station C. Zeniths angles measured at B when sighting on the HI on C were 284°11'54" and 75°49'24".

Station	Departure Coordinate	Latitude Coordinate
B	9872.65	9907.11
C	9744.54	9758.71

8. Slope distance from A to B = 114.5 links. Horizontal distance AB = 1 chain Slope distance from A to C = 223.0 links. Horizontal distance AC = 2 chains Bearing of AB = bearing of AC
B lies at a lower elevation than both A and C. Find the slope from B to C in percent.

AREAS

Often one of the objectives of the survey is to find the area of the tract of land which has been traversed. The area of an irregular shaped figure is not difficult to determine mathematically if there are few sides to the traverse. If the figures are regular shaped, formulae may be used to find the area as shown in Figure 5-1. The area may also be determined with a planimeter after the traverse is drawn to scale but this method is not accurate.

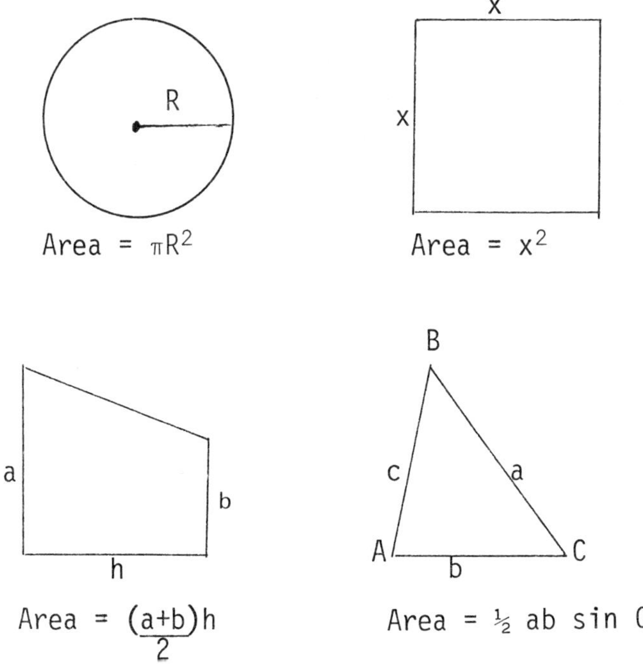

Figure 5-1. Area of regular shaped figures.

Areas of irregular tracts may often be computed mathematically by dividing the tract into triangles. The area of each triangle is computed and the total area determined by adding all of the areas together. Figure 5-2 shows the creation of triangles for the traverse using data from a previous section of the text.

Course	Distance	Bearing
AB	120.61'	N 86°00'W
BC	114.74'	N 56°50'W
CD	191.50'	N 42°00'E
DE	156.81'	S 68°50'E
EA	167.92'	S 21°00'W

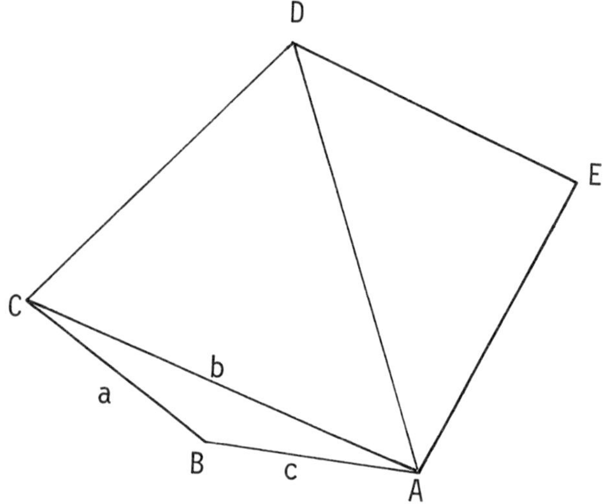

Figure 5-2. Area of traverse by dividing into triangles.

The areas of triangle ABC and triangle ADE can be computed using the formula Area = ½ a • b • sin ∡ C. Hence, the areas are as follows:

$$\text{Area ABC} = \frac{120.61 \times 114.74}{2} \times \sin 150°50' = 3372.18$$

$$\text{Area ADE} = \frac{156.81 \times 167.92}{2} \times \sin 89°50' = 13165.71$$

In triangle ACD, only the length of CD is known so the above formula can not be used to solve for the area until more data is computed. There are four different solutions that could be used: (1) Solve for the area using three sides of the triangle. (2), (3) and (4) Solve for the area using two sides and the included angle where the angles CAD, ADC and DCA are the interior angles of the triangle. Angle ACD could be computed by first solving for angle BCA using the law of tangents as follows:

$$\text{Tangent}\left[\frac{∡\,C - ∡\,A}{2}\right] = \frac{c - a}{c + a} \times \text{tangent}\left[\frac{∡\,C + ∡\,A}{2}\right]$$

$$∡\,ABC = 150°50', \qquad ∡\,A + ∡\,B + ∡\,C = 180°$$

$$∡\,A + ∡\,C = 180° - 150°50'$$

$$\left[\frac{∡\,A + ∡\,C}{2}\right] = 14°35'$$

$$c = 120.61$$

$$a = 114.74$$

$$\text{Tangent}\left[\frac{\angle C - \angle A}{2}\right] = \frac{5.87}{235.35} \times \text{tangent } 14°35'$$

$$= 0.006489$$

(1) $\quad \left[\frac{\angle C - \angle A}{2}\right] = 0°22'18''$

(2) $\quad \left[\frac{\angle C + \angle A}{2}\right] = 14°35'$

Adding (1) and (2) $\quad \angle ACB = 14°57'18''$

Subtracting (1) from (2) $\quad \angle CAB = 14°12'42''$

Side b is computed by applying the law of sines.

$$\frac{b}{\sin 150°50'} = \frac{120.61}{\sin 14°57'18''} = \frac{114.74}{\sin 14°12'42''}$$

b = 227.77 feet

$\angle BCD = 180° - [56°50' + 42°00'] = 81°10'$

$\angle ACD = 180° - (98°50' + 14°57'18'') = 66°12'42''$

Area of triangle ACD can now be computed using two sides and the included angle.

Area of ACD

$$= \frac{227.77 \times 191.50}{2} \times \sin 66°12'42'' = 19956.11$$

Area of ABC $\qquad = 3372.18$

Area of ADE $\qquad = 13165.71$

Total area $\qquad = 36494$

Another solution would be to first solve for side b in triangle ABC by using the law of cosines. The law of cosines, like the law of tangents, is used when two sides and the included angle of a triangle are known.

Law of cosines:

$$b = \sqrt{a^2 + c^2 - 2ac \cos \beta}$$

$$c = 120.61'$$

$$a = 114.74'$$

$$\angle B = 150°50'$$

$$= \sqrt{(114.74)^2 + (120.61)^2 - 2(114.74)(120.61)(\cos 150°50')}$$

$$= \sqrt{13165.27 + 14546.77 - (-24168.22)}$$

$$= \sqrt{51880.26}$$

$$= +227.77$$

The next step, having solved for side b, would be to solve for $\angle$ ACD using the law of sines and then proceed as before when the law of tangents was used to solve for $\angle$ BCA. It is to be noted that if the included angle used in the law of cosines is greater than $90°$ the cosine function is negative and the value of the term $2(ac)(\cos \beta)$ is also negative.

Using the law of cosines:

$$AD = \sqrt{(156.81)^2 + (167.92)^2 - 2(156.81)(167.92)(\cos 89°50')}$$

$$= \sqrt{24589.38 + 28197.13 - 153.19}$$

$$= \sqrt{52633.32}$$

$$= 229.42$$

Computations for the area of triangle ACD using three sides is as follows:

$$AC = 227.77' \qquad S = \frac{a + b + c}{2}$$

$$AD = 229.42' \qquad S = \frac{648.69}{2}$$

$$CD = 191.50' \qquad S = 324.345'$$

$$\text{Area ACD} = \sqrt{S(S-a)(S-b)(S-c)}$$

$$= \sqrt{324.345(96.575)(94.925)(132.845)}$$

$$= \sqrt{395000589}$$

$$= 19874.62^*$$

*The discrepancy of 81 square feet between the areas found using three sides of the triangle and two sides and the included angle is due to the use of data which had not yet been corrected for the error that existed in the closure of the traverse.

Sometimes it is necessary to solve for the area bound by a curved boundary such as a road or a stream. In these situations, it is best to run a traverse near the irregular boundary and measure at regular intervals right angle offsets to the irregular boundary. The area is computed using the trapezoidal rule. It is not the most accurate method because it considers the curved boundary to be a straight line between the offsets but if the offsets are made at small intervals this error will be minimal considering the inaccuracies of measuring the offsets.

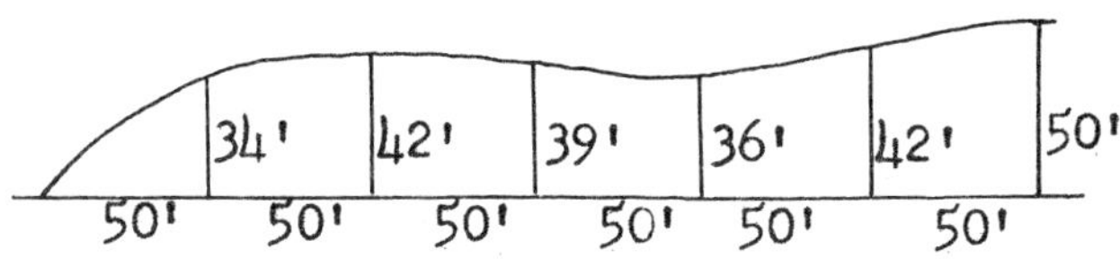

Figure 5-3. Area of figure bound by curved boundary.

If the offsets are made at regular intervals of h as shown in Figure 5-3, then the total area bounded by a straight line and a curve could be considered to be the summation of a series of trapezoids whose individual areas are $[(a+b)/2] \times h$. If the first offset were n_0 and the last offset were n_n, the sum of the areas of the trapezoids would be

$$[\frac{n_0+n_1}{2}]h + [\frac{n_1+n_2}{2}]h + [\frac{n_2+n_3}{2}]h + ... [\frac{n_{n-1}+n_n}{2}]h$$

which would equal the total area if there were n_n offsets. The sum of all of the individual trapezoids is better expressed by the following equation;

$$\text{Area} = h\left[\frac{n_0 + n_n}{2} + n_1 + n_2 + n_3 ... n_{n-1}\right]$$

The area of the traverse in Figure 5-3 would then equal

$$50\left[\frac{0 + 50}{2} + 34 + 42 + 39 + 36 + 42\right] = 50[218] = 10,900 \text{ sq.ft.}$$

The area of larger irregular tracts can be computed mathematically by other methods depending on what type of data is available. If the coordinates of each point are known, the coordinate method is used. A derivation of this method is shown for Figure 5-4. It is also assumed that the figure for which the area is to be computed lies wholly in the first quadrant so all of the coordinate values are positive.

Find: Area of ABCDA.

Solution: Area of ABCDA = area bBCc + area cCDd – area bBAa – area aADa.

The areas shown are trapezoids whose parallel sides are m_1, m_2, m_3 and m_4. The values m_1, m_2, m_3 and m_4 are the perpendicular distances from the points A, B, C and D to the reference meridian. This meridian distance is also equal to the departure coordinate of each point respectively. The lines p_1, p_2, p_3 and p_4 represent the perpendicular distances to the reference parallel and these distances are equal to the latitude coordinates of points A, B, C and D. The area of ABCDA is the sum of the areas of trapezoids whose areas in terms of parallel sides and perpendicular distances between the parallel sides is as follows,

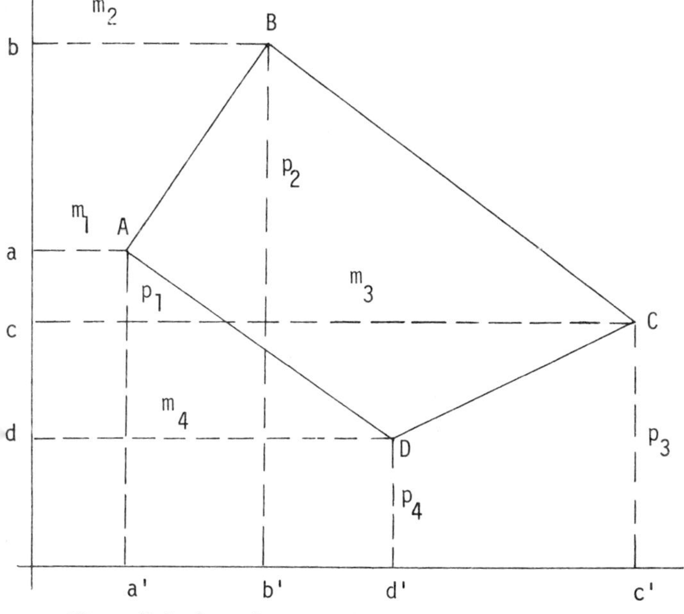

Figure 5-4. Coordinate method of solving for area.

$$\text{Area ABCDA} = \left[\frac{m_2 + m_3}{2}\right](p_2 - p_3) + \left[\frac{m_3 + m_4}{2}\right](p_3 - p_4)$$

$$- \left[\frac{m_2 + m_1}{2}\right](p_2 - p_1) - \left[\frac{m_1 + m_4}{2}\right](p_1 - p_4)$$

In simplifying this equation, it should be noted that the common denominator of 2 is removed and that results in solving for double area or twice the area desired.

$$2 \times \text{Area ABCDA} = (m_2+m_3)(p_2-p_3) + (m_3+m_4)(p_3-p_4)$$
$$- (m_2+m_1)(p_2-p_1) - (m_1+m_4)(p_1-p_4)$$

$$2 \times \text{Area ABCDA} = m_2 p_2 - m_2 p_3 + m_3 p_2 - m_3 p_3$$
$$+ m_3 p_3 - m_3 p_4 + m_4 p_3 - m_4 p_4$$
$$- m_1 p_2 + m_1 p_1 - m_2 p_2 + m_2 p_1$$
$$- m_1 p_1 + m_1 p_4 - m_4 p_1 + m_4 p_4$$

Simplified:

$$2 \times \text{Area ABCDA} = - m_2 p_3 + m_3 p_2 - m_3 p_4 + m_4 p_3$$
$$- m_1 p_2 + m_2 p_1 + m_1 p_4 - m_4 p_1$$

Rearranged:

$$2 \times \text{Area ABCDA} = - m_1 p_2 + m_1 p_4 - m_2 p_3 + m_2 p_1$$
$$+ m_3 p_2 - m_3 p_4 - m_4 p_1 + m_4 p_3$$

Changing the signs which has no effect on the area, the above equations becomes:

$$2 \times \text{Area ABCDA} = m_1 p_2 + m_2 p_3 + m_3 p_4 + m_4 p_1 - m_2 p_1$$
$$- m_3 p_2 - m_4 p_3 - m_1 p_4$$

A simpler method of solving for areas using coordinates, remembering that the m value is the departure coordinate of the point and the p value is the latitude coordinate, is to arrange the values as follows.

$$\left[\dfrac{DC}{LC}\right] \quad \dfrac{m_1}{p_1} \ \dfrac{m_2}{p_2} \ \dfrac{m_3}{p_3} \ \dfrac{m_4}{p_4} \ \dfrac{m_1}{p_1}^{*}$$

*The coordinates of the first point are repeated.

Having arranged the values as shown above, cross multiplication is performed, i.e., m_1 is multiplied by p_2 and m_2 is multiplied by p_3 and so on. These products are assumed to have positive values. Cross multiplication is performed in the opposite direction and it is assumed that these products have negative values. Therefore,

$$\dfrac{m_1}{p_1} \qquad \dfrac{m_2}{p_2} \qquad \dfrac{m_3}{p_3} \qquad \dfrac{m_4}{p_4} \qquad \dfrac{m_1}{p_1}$$

becomes $m_1 p_2 + m_2 p_3 + m_3 p_4 + m_4 p_1 - m_2 p_1 - m_3 p_2 - m_4 p_3 - m_1 p_4$ which is the same form that was obtained by solving for the areas of the individual trapezoids. The area may also be found by placing all the latitude coordinates in the numerator and all the departure coordinates in the denominator.

The difference between the sum of the positive multiplications and the sum of the negative multiplications is obtained; this, disregarding the sign, is the double area. The area of Figure 5-4 using the assumed coordinate values would be computed as follows.

Station	Latitude Coordinate	Departure Coordinate
A	100	30
B	170	80
C	75	200
D	40	120

$$\frac{30}{100} \qquad \frac{80}{170} \qquad \frac{200}{75} \qquad \frac{120}{40} \qquad \frac{30}{100}$$

$$+$$

(30 X 170)	5100	8000 (80 X 100)
(80 X 75)	6000	34000 (170 X 200)
(200 X 40)	8000	9000 (75 X 120)
(120 X 100)	12000	1200 (30 X 40)
	Σ=+31100	Σ=−52200

2 X Area = +31100 – 52200

2 X Area = – 21100 (negative sign is now discarded)

2 X Area = ABCDA = 21100

Area ABCDA = 10,550 (square units)

A second method of solving for the area of irregular figures is known as the DMD or Double Meridian Distance method and is used if the latitudes and departures of the individual courses are known. It is necessary to understand certain definitions before proceeding further.

1. The meridian distance of a point is the perpendicular distance from the reference meridian to that point.

2. The double meridian distance of a line is the sum of the meridian distances of its two extremities.

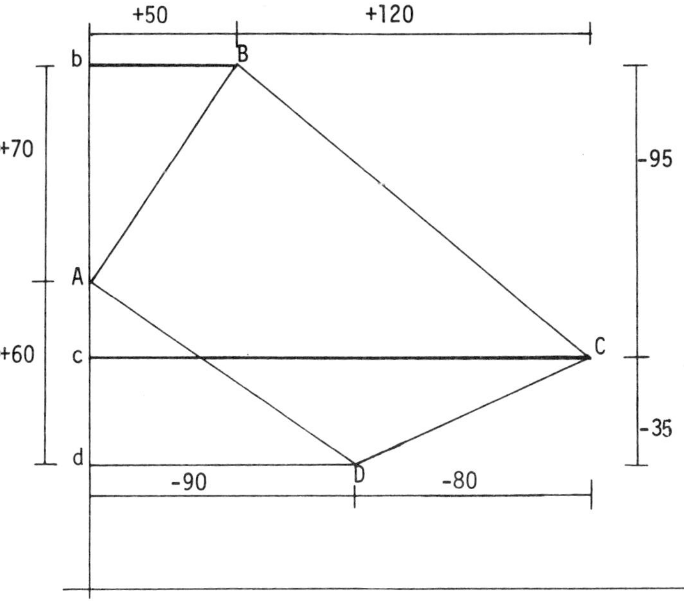

Figure 5-5. Double meridian distance method of solving for area.

In Figure 5-5, it can be seen that the double meridian distance of line BC in the traverse ABCDA is = bB + Cc. It should also be observed that these two sides are parallel and form two sides of a trapezoid bBCc whose area is = (bB+Cc)/2 × the perpendicular distance between the parallel sides. This distance is the latitude of the line BC. Hence, 2 × area bBCc = (bB+Cc) × latitude of BC. However the DMD of line BC = bB + Cc so that 2 × area bBCc = DMD of BC × latitude of BC.

The figure ABCDA is made up of several trapezoids and the area is found by combining all of the areas. The sign ($\pm$) of the area of each trapezoid is determined by the sign of the latitude of the course. The sketch in Figure 5-5 has the reference meridian passing through the extremity of line AB. The double meridian distance of each line would be as follows:

Line	DMD	DMD
AB	0 + bB	0 + 50 = 50
BC	bB + cC	50 + 170 = 220
CD	cC + dD	170 + 90 = 260
DA	dD + 0	90 + 0 = 90

Three rules must be followed to determine the double meridian distance (DMD) of each course when the coordinates are not known.

1. The DMD of the first course is the departure of that course. (This is assuming that the reference meridian passes through one extremity of that course.)

2. The DMD of any other course is equal to the DMD of the preceding course + the departure of the preceding course + the departure of the course itself.

3. The DMD of the last course is equal to the departure of the last course but with the opposite sign. (The DMD of the last course should be computed and then compared with the departure as a check. If the traverse has not been balanced; i.e., the Σ departures is not equal to zero, the DMD of the last course will differ from the departure of the last course by an amount that is numerically equal to two times the Σ departures.)

Figure 5-6 shows the areas of the trapezoids that combine to give the total area of ABCDA computed in the manner shown below.

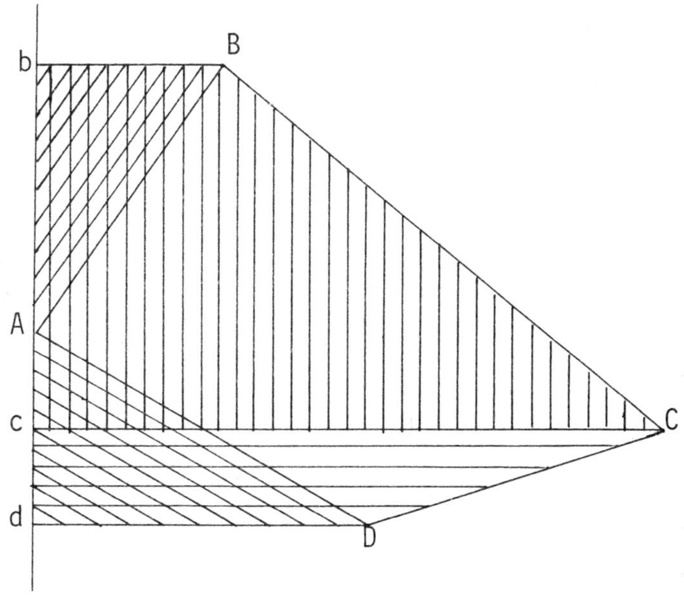

Figure 5-6. Area by DMD method (+ and – areas).

Line	AB	BC	CD	DA
Latitude	+70	-95	-35	+60(Σlat.=0)
Departure	+50	+120	-80	-90(Σdep.=0)
DMD	+50	+220	260	+90
Double Area	+3500	-20900	-9100	+5400

D.A. = +8900 – 30000 = –21100

D.A. – 21100 (negative sign discarded)

Area = 21100 ÷ 2 = 10,550 (square units)

Many different problems dealing with the subdivision of land can be solved easily by the use of latitudes and departures, coordinates and trigonometry. The most common types of problems are:

1. Cutting off an area by running a straight line between two points.

2. Cutting off a given area by running a straight line through a given point.

3. Running a straight line from a point on a given bearing.

4. Cutting off a given area by running a line on a given bearing.

The traverses shown in Figure 5-7 with the following coordinates will be used to illustrate all four types of problems. The origin was arbitrarily placed at Station E so that there is a negative departure coordinate to consider.

Station	Latitude Coordinate	Departure Coordinate
A	+50	-30
B	+100	+25
C	+60	+100
D	+10	+80
E	0	0

1. Cut off an area with a straight line from A to C. Since all of the coordinates are known, it is easy to compute the area of ABC by the coordinate method.

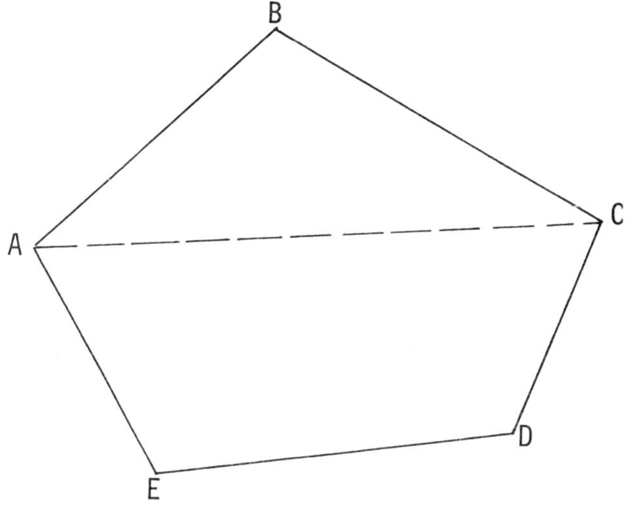

Figure 5-7. Dividing an area with a line between two points.

50	100	60	50
-30	25	100	-30

+	-
1250	-3000
10000	1500
-1800	5000
$\Sigma = +9450$	$\Sigma = -3500$

Double Area = 9450 – 3500 = 5950

Area ABC = $\dfrac{5950}{2}$ = 2,975 square units

2. Divide the area into two equal parts by running a line through station A. Compute the length and bearing of this line and check the results.

The total area and the half area of the traverse are first computed by the coordinate method.

Station	A	B	C	D	E	A
Dep. Coord.	-30	25	100	80	0	-30
Lat. Coord.	50	100	60	10	0	-30

+	-
-3000	1250
1500	10000
1000	4800
0	0
0	0
Σ = -500	Σ = -16050

Double Area = 16550

Area ABCDEA = 8275 square units

½ Area ABCDEA = 4137.50 square units

The line from A to C cut off only 2975 square units, triangle ABC, in the previous problem and therefore is not the line that divides the area into two equal parts. So now assume that the line from A to H shown in Figure 5-8 is the desired line and that the area of ABCHA = 4137.5 square units. The area of ACH must then be equal to the area ABCHA - area ABC or 4137.50 - 2975, or 1162.50 square units. In order to find the length and bearing of a line, as is required for this problem, it is necessary to know the coordinates at both ends of the line. If the coordinates of a point at one end of the line are unknown, they can be determined if the latitude and the departure of a connecting line from a point whose coordinates are known can be computed. Therefore, the coordinates of H could be computed if the distance CH were known since the bearing of CH can be computed from the known coordinates at C and D. In the triangle ACH, the length and bearing of AC can be determined and the angle ACH computed.

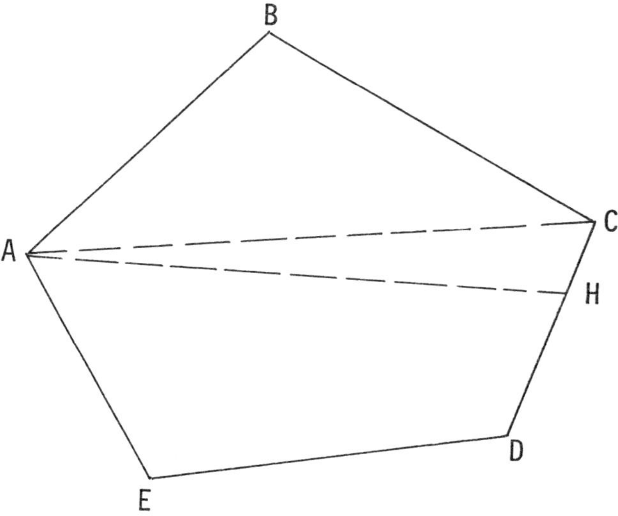

Figure 5-8. Dividing an area into two equal parts by running a line through a given point.

$$\text{Bearing AC} = \tan^{-1} \frac{130}{10} = \text{N } 85°36'05''\text{E}$$

$$\text{Distance AC} = \frac{130}{\sin 85°36'05''} = 130.38$$

Bearing CD = Bearing CH = $\tan^{-1} \dfrac{-20}{-50}$ = S 21°48′05″W

$\measuredangle$ ACH = 85°36′05″ − 21°48′05″ = 63.79988°

(Angle changed to decimal for calculator)

The area of a triangle can be expressed as ½ ab sin $\measuredangle$ C. Hence, the area of ACH = 1162.50 = $\dfrac{AC \times CH \times \sin \measuredangle ACH}{2}$.

In this equation, the only unknown quantity is CH. Solving for CH which = $\dfrac{2 \times 1162.50}{130.38' \times \sin 63.79988°}$ CH is found to be 19.8738′.

Latitude CH = 19.8738 × cos 21.8014°
= −18.4524 since bearing of CD is SW

Departure CH = 19.8738 × sin 21.8014°
= −7.3809 since bearing of CD is SW

Lat. Coord. H = Lat. Coord. C + Latitude CH
= 60 − 18.4524 = 41.5476

Dep. Coord. H = Dep. Coord. C + Departure CH
= 100 − 7.3809 = 92.6191

A check is made to determine whether or not the line AH divides the area into two equal parts by solving for the area AHDEA using coordinates.

Station	A	H	D	E	A
Dep. Coord.	−30	92.6191	80	0	−30
Lat. Coord.	50	41.5476	10	0	50

$$
\begin{array}{cc}
+ & - \\
-1246.43 & 4630.95 \\
926.19 & 3323.81 \\
0 & 0 \\
\underline{\hspace{2em}0} & \underline{\hspace{2em}0} \\
\Sigma = -320.24 & \Sigma = -7954.76
\end{array}
$$

Double Area = 8275.00

Area AHDEA = 4134.50

$$
\text{Bearing AH} = \tan^{-1}\frac{122.6191}{-8.4524} = \text{S } 86°03'24''\text{E}
$$

$$
\text{Distance AH} = \sqrt{(122.6191)^2 + (-8.4524)^2}
$$
$$
= 122.91'
$$

A second method of solving for the length and bearing of a line which divides the traverse into two equal parts is to solve for the coordinates of H using simultaneous equations. The latitude coordinate is designated as Y and the departure coordinate is designated as X. Using the known coordinates of the other stations, two equations can be written which will express the area of ABCHA and AHDEA with two unknowns Y and X. Using the coordinate method, the coordinates are arranged in order as though both areas were being traversed in a counter-clockwise direction.

Area of upper half or AHCBA:

Station	A	H	C	B	A
Dep. Coord.	-30	X	100	25	-30
Lat. Coord.	50	Y	60	100	50

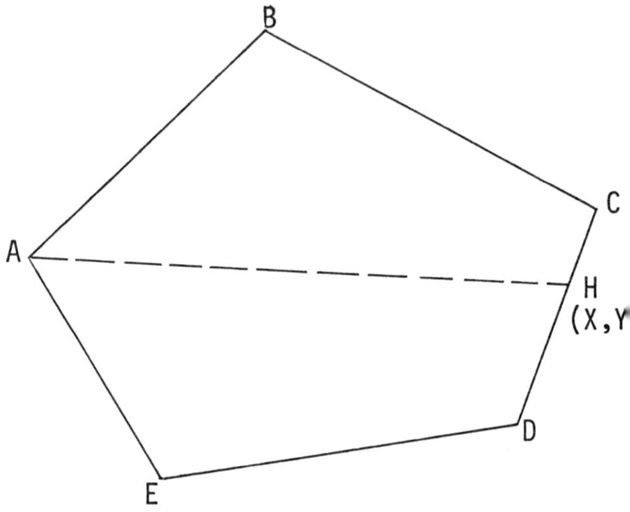

Figure 5-9. Use of simultaneous equations to find length
and bearing of line.

+	−
−30 Y	50 X
+60 X	100 Y
+10000	1500
+1250	−3000

$$-30Y - 100Y + 60X - 50X + 11250 + 3000$$

$$- 1500 = 2 \cdot Area$$

$$-130Y + 10X + 12750 = 8275$$

$$(1) \quad 130Y - 10X = 4475$$

Area of lower half or AEDHA:

Station	A	E	D	H	A
Dep. Coord.	-30	0	80	X	-30
Lat. Coord.	50	0	10	Y	50

+	-
0	0
0	0
80 Y	10 X
50 X	-30 Y

$80Y + 50X - 10X + 30Y = 2 \cdot \text{Area}$

(2) $110Y + 40X = 8275$

Multiply (1) $\times$ 4 $= 520Y - 40X = 17900$
          (2) $= 110Y + 40X = 8275$

Adding: $630Y = 26175$
          $Y = 41.5476$

Substituting: $X = 92.619$

These coordinate values are identical to those found using trigonometry.

Figure 5-10 illustrates the division of a traverse by a line running in a given direction and passing through a given point.

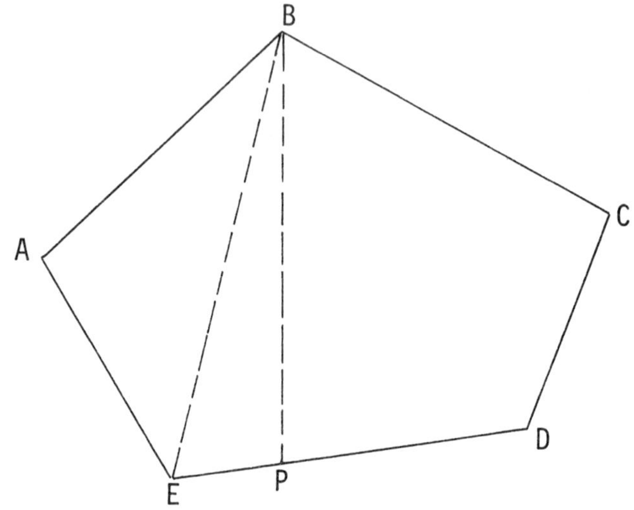

Figure 5-10. Cutting off an area with a line in a given direction through a given point.

3. Determine the area cut off by a line through station B on a bearing of S 1°00'E. Again this is a problem of finding the coordinates of P by solving for the distance DP in the triangle BDP or the distance EP in the triangle BEP. The following results were obtained.

Bearing BP = S 1°00'E   ∡ BPE = 96.125°
Bearing BE = S 14.03624°W   ∡ PEB = 68.8386°
Distance BE = 103.9776'   ∡ EBP = 15.0362°
Bearing DE = S 82.8798°W   Distance EP = 26.8949'
Latitude EP = 3.3358   Lat. Coord. P = 3.3358
Departure EP = 26.6872   Dep. Coord. P = 26.6872

The area of BCDPB (4857.5) + area of BPEAB (3417.5) = 8275 which is the total area of ABCDEA thus proving that the coordinates of P are correct.

A simpler solution would be to solve for the coordinates of station P using simultaneous equations.

Line DE can be expressed as follows since the coordinates of the end points of the line are known.

$$\frac{Y-Y_1}{Y_2-Y_1} = \frac{X-X_1}{X_2-X_1}$$

$$\frac{Y-0}{10-0} = \frac{X-0}{80-0}$$

(1) $80Y - 10X = 0$

Line BP can be expressed in terms of the slope since the coordinates of only one point and the bearing of the line are known.

$$(Y-Y_1) = \cot \beta \, (X-X_1)$$

$$Y - 100 = \cot 1° (X-25)$$

$$Y - 100 = -57.28996(X-25)$$

(2) $Y + 57.28996X = 1532.249$

Dividing (1) by 80: $Y - 0.1250X = 0$

Subtracting: $57.4149X = 1532.2490$

$X = 26.6873 = $ Dep. Coord. of P

Substituting: $Y = 3.3359 = $ Lat. Coord. of P

These values are equal to the coordinates of P previously computed using trigonometry.

4. Cutting off a given area by running a line in a given direction is the most complicated of all types of division of an area. In Figure 5-11 it is desired to divide the traverse into two equal areas with a line that is parallel to side AB. This, then, becomes a problem of finding the coordinates at both extremities of a line.

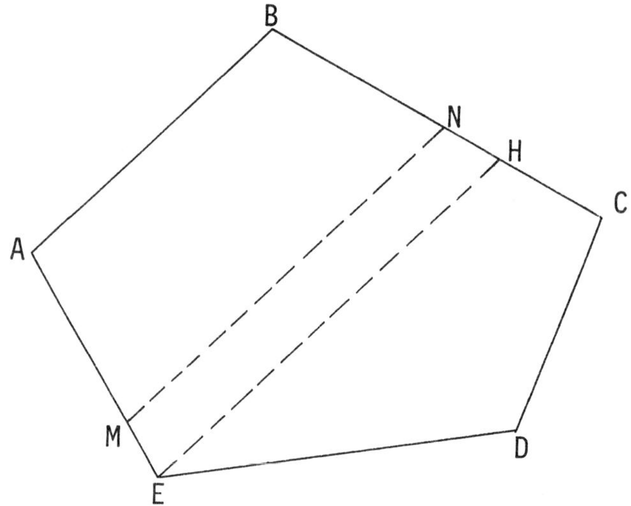

Figure 5-11. Dividing an area into two equal parts with a line running in a given direction.

Area of ABCDEA = 8275 square units and ½ ABCDEA
                = 4137.5 square units.

The first step is to run a line parallel to side AB through station E and determine if this line will divide the area in half. The area of ABHEA can only be found after computing the coordinates of H which is done using simultaneous equations.

Bearing AB = $\tan^{-1} \dfrac{55}{50}$ = N 47.7263 E

Equation of line EH $(Y-0) = \cot 47.7263^\circ (X-0)$

(1)   Y = 0.90909X

Equation of line BC $\dfrac{Y-100}{60-100} = \dfrac{X-25}{100-25}$

(2)   75Y + 40X = 8500

Divide (2) by 75: Y + 0.5333X = 113.333

Multiply (1) by –1: –Y + 0.90909X = 0

Adding: 1.44241X = 113.333
                    X = 78.5720 = Dep. Coord. H
                    Y = 71.4286 = Lat. Coord. H

Using the coordinates of H in the coordinate method for solving for areas, the area of ABHEA is found to be 5160.7139 square units. However, ½ of the total area of the traverse is only 4137.50. Therefore, the line EH does not divide the traverse into two equal parts and the area ABHEA must be reduced by 1023.2139 square units

(5160.7139 – 4137.50). The area of the trapezoid EMNH which is created by a line parallel to EH and hence parallel to AB is equal to 1023.2139 square units. The area of EMNH = [(EH+MN)/2]·X. (The term X used here denotes an unknown value and not a Dep. Coord.) Since there are two unknowns in this equation, MN and X, the distance MN is then expressed in terms of EH and X leaving only one unknown term.

$$\text{Distance EH} = \sqrt{(78.5714)^2+(71.4286)^2} = 106.1862'$$

$$\text{Bearing EA} = \tan^{-1}\frac{-30}{50} = \text{N } 30.963757°\text{W}$$

$$\text{Bearing CB} = \tan^{-1}\frac{-75}{40} = \text{N } 61.9275°\text{W}$$

Figure 5-12 shows that the rectangular area obtained by multiplying EH times X (X is perpendicular to EH) exceeds the area of the trapezoid EMNH by an amount equal to the two small triangles created by the lines AE and BE. The area of the trapezoid then becomes:

$$\text{EH}\cdot\text{X} - \left[\frac{\text{X}\cdot\text{X}\tan\theta}{2} + \frac{\text{X}\cdot\text{X}\tan\phi}{2}\right] = 1023.2139 \text{ square units}$$

$$\text{EH}\cdot\text{X} - \frac{\text{X}^2}{2}(\tan\theta + \tan\phi)$$

$$\theta = (90°-47.7263°) - 30.96375° = 11.3099°$$

$$\phi = 61.9275° - (180° - 47.7263°) + 90° = 19.6538°$$

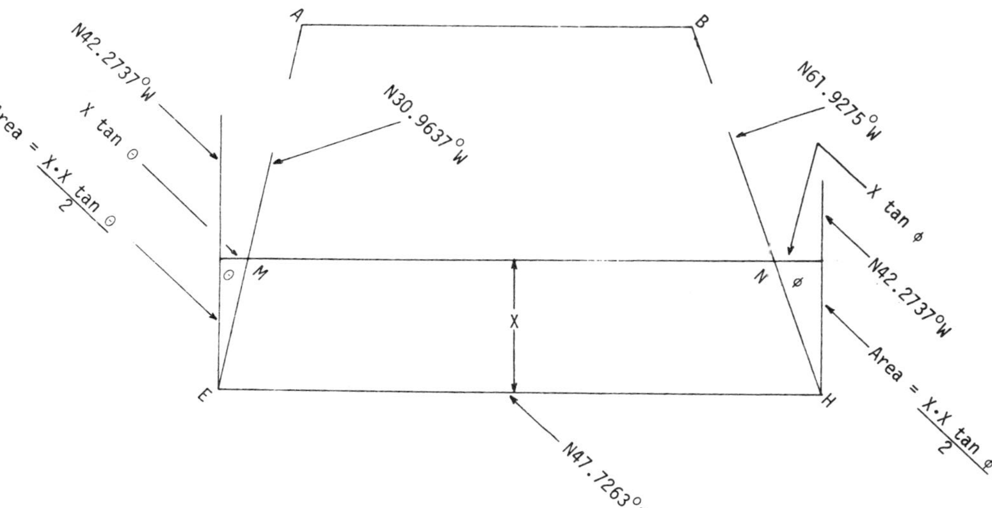

Figure 5-12. Enlargement of Figure 5-11.

154

Since $1023.2139 =$

$$= EH \cdot X - \frac{X^2}{2} \text{ [tangent } 11.3099° + \text{ tangent } 19.6539°]$$

is a quadratic equation, it may be solved by using the formula

$$X = \frac{-b \pm \sqrt{b^2 - 4ac}}{2a}$$

or by completing the square. Substituting and simplifying the equation becomes:

$$1023.2139 = 106.1862 \cdot X - \frac{X^2}{2} (0.20000 + 0.357142)$$

$$1023.2138 = 106.1862 \cdot X - 0.278571 \ X^2 \quad \text{or}$$

$$0.278571 \ X^2 - 106.1862X + 1023.2138 = 0$$

Using the formula to solve for X the values of a and b are 0.278571 and –106.1862 respectively.

$$X = \frac{106.1862 \pm \sqrt{(106.1862)^2 - (4)(0.278571)(1023.2138}}{(2)(0.278571)}$$

$$= \frac{106.1862 \pm \sqrt{10135.36}}{0.557142}$$

$$= \frac{5.511675}{0.557142} = 9.892765'$$

The method of completing the square is as follows:

1. Write the equation so that the coefficient of X = 1.

2. Transfer the constant (c) to the right hand side of the equation.

3. Divide the coefficient of the X term (b) by 2, square and add to both sides of the equation.

4. The left side of the equation is now a perfect square, i.e., $(X-(b/2))^2$ and is solved for X by taking the square root of both sides of the equation.

The original equation was

$$0.278571X^2 - 106.1862X + 1023.2138 = 0$$

and this becomes:

$$X^2 - 381.1818X + 36324.8944 = 36324.8944 - 3673.0808$$

$$(X-190.5909)^2 = 32651.8135$$

$$X - 190.5909 = -180.698128 \text{ (negative value used)}$$

$$X = 9.8927 \text{ (same value as obtained with the formula)}$$

$$\text{Distance EM} = \frac{9.8927}{\cos 11.30993°}$$

$$EM = 10.0886'$$

Latitude EM = 10.0886 $\times$ cos 30.9637° = 8.6509

Departure EM = 10.0886 $\times$ sin 30.9637° = -5.1905

Lat. Coord. M = 0 + 8.6509 = 8.6509

Dep. Coord. M = 0 - 5.1905 = -5.1905

Distance HN $= \dfrac{9.8927}{\cos 19.6538^\circ} = 10.5046'$

Latitude HN $= 10.5046 \times \cos 61.9275^\circ = +4.9433$

Departure HN $= 10.5046 \times \sin 61.9275^\circ = -9.2688$

Lat. EH $= 106.1862 \times \cos 47.7263^\circ = 71.4285$

Dep. EH $= 106.1862 \times \sin 47.7263^\circ = 78.5714$

Lat. Coord. N $= 71.4285 + 4.9433 = 76.3718$

Dep. Coord. N $= 78.5714 - 9.2688 = 69.3026$

Area EMNCDE (Check)

Station	E	M	N	C	D	E
Dep. Coord.	0	-5.1905	69.3026	100	80	0
Lat. Coord.	0	8.6509	76.3718	60	10	0

+	-
-396.4078	599.5298
4158.1560	7637.1800
1000.0000	4800.0000
$\Sigma = 4761.7482$	$\Sigma = 13036.7098$

Double Area $= 8274.9616$

CHECK   Area $= 4137.48 = \tfrac{1}{2}$ Area $= 4137.50$

1.   The land between 1st and 2nd Street and Oak and Marlow is to be divided into 3 lots with the boundaries running parallel to Oak Street.   The east and west lots each have 1/5 of the total area. All   streets   are   60.0   feet   wide.

FIND: Total area of lots in the block in square feet. The length of frontage on Marlow Street of the easterly most lot.

The frontage of the most westerly lot on 2nd. Street. The frontage on 1st Street of the middle lot.

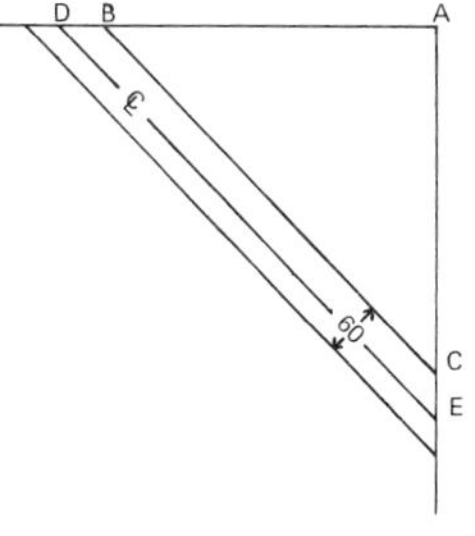

2. Bearing AD = N 87°30′W
   Bearing AC = S 4°00′E
   Bearing DE = S 48°00′E
   Distance DE = 710 feet = ₵ of road
   Road width - 60.0 feet
   Find the area of ABC.

3.   Find the length and bearing of a line through the midpoint of line AB which will divide the area of the tapezoid ABCD into equal areas.

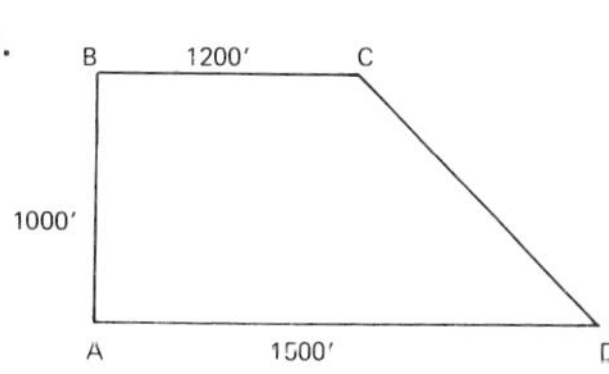

4. Find the area of ABCDEA by the DMD, DPD and co-ordinate methods.

Course	Latitude	Departure
AB	+90	+40
BC	−110	+10
CD	+40	+60
DE	−120	−80
EA	+100	−30

5.

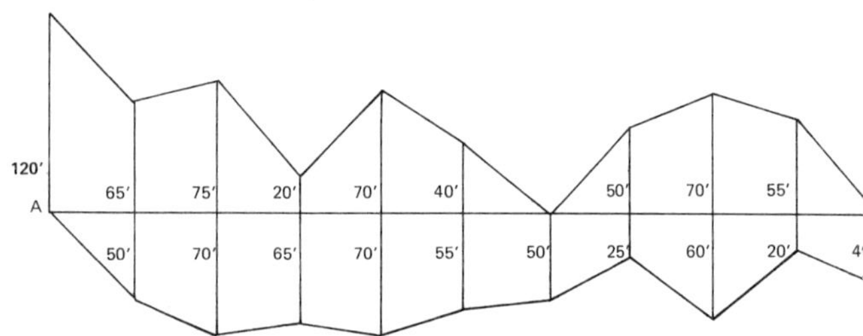

Find the area of the figure above. Offsets are made at right angles to the line AB at intervals of 50.0 feet.

6. Station/Slope Distance/Deflection Angle/Vertical Angle

	C		
	148.6 feet	114°45'00''R	+2°30'
B		57°22'30''R	+2°27'
	168.7 feet		+6°08'
A			+6°07'

Find the area of ABC and the distance AC

STADIA

Stadia is a method of measuring distances, both horizontal and vertical by rapid indirect methods. A transit or theodolite and a leveling rod are used. The process is called tacheometry and is one in which the distance between the transit or theodolite and the leveling rod, also called a stadia rod, is computed by observing an interval on the rod between two horizontal hair lines in the telescope. The level rod shown in Figure 6-1 is calibrated in feet, tenths and hundredths of feet. In stadia, the intersection of the image of the two hair lines (upper and lower) on the rod, held vertically so it is perpendicular to the line of sight, is observed. The interval between the hairlines is called the stadia interval R or the rod intercept and is read to the nearest one hundredth of a foot. The spacing and the width of each mark on the level rod in Figure 6-1 is 1/100 of a foot. Failure to read the rod to the nearest 1/100 of a foot could result in an error in horizontal distance of two feet per hundred if both the upper and lower hair lines were incorrectly read.

In Figure 6-2, the upper and lower hair lines are represented by b and a and the distance between them is i which is also equal to a'b'. F is the principal focus and f is the focal length of the instrument. Hence, by similar triangles $f/i = d/R$. The distance from the rod to the principal focus is therefore $= d = (f/i)R = KR$. For almost all instruments, the value of $f/i = K = 100$. This is to say that the i value is such that the upper and lower hair lines will intersect at intervals of 1.00 foot on the rod when the distance d = 100 feet. The distance from the rod to the instrument $= d + C$ in which the distance C is a constant and is equal to the

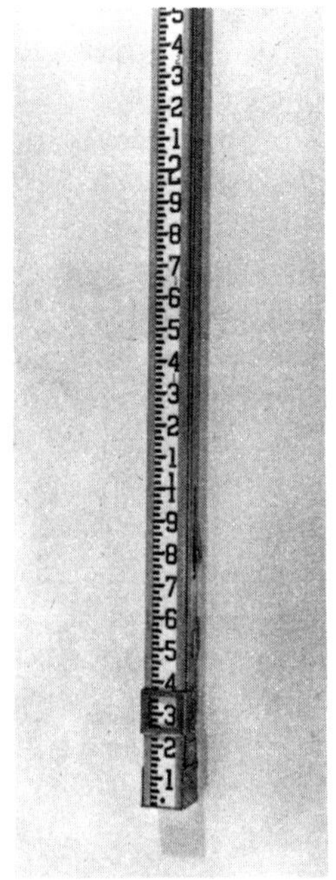

Figure 6-1. Level rod (courtesy Lietz Co.).

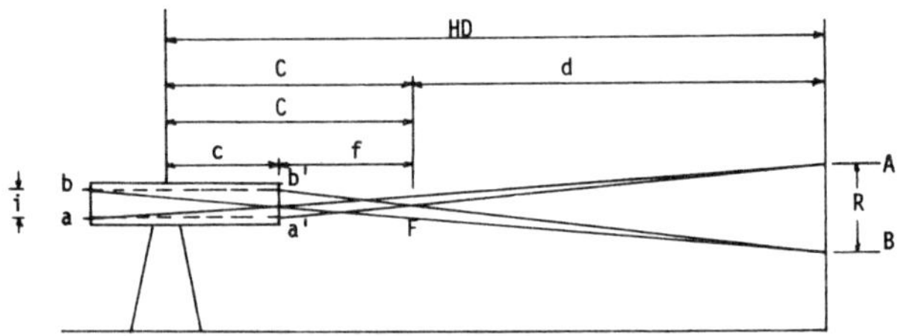

Figure 6-2. Stadia principal.

distance from the center of the instrument to the principal focus. Only when using an external focusing type of transit does the value of C have to be considered and then the HD = d + C = KR + C. In an internal focusing type of transit, C is equal to zero and can be ignored and the distance from the instrument to the rod then becomes KR.

It is assumed that the line of sight is perpendicular to the rod when using the equation d = KR. This is not the case when considering inclined shots where the vertical angle may be + or − because the rod is always held in a vertical position. Figure 6-3 illustrates an inclined line of sight. For all practical purposes the line of sight is perpendicular to the rod if the rod were in the A′B′ position but with the rod vertical the rod intercept is really AB. The rod intercept A′B′ therefore has to be computed and would equal the rod intercept AB $\times$ cosine α where α is the vertical angle. The slope distance (SD) = K(A′B′) + C but since the reading of the upper stadia hair − the reading of the lower stadia hair (UH-LH) = R, which equals AB, the slope distance becomes K $\times$ AB $\times$ cosine α + C and if C = 0, the SD = 100AB cos α. The horizontal distance which equals the slope distance $\times$ cos α now equals K $\times$ R $\times$ cos$^2\alpha$ + C $\times$ cos α.

The difference in elevation which = SD $\times$ sin α becomes K $\cdot$ R $\cdot$ cos α $\cdot$ sin α + C $\cdot$ sin α. It should be noted again that in an internal focusing type of transit which has a C value of 0, the Horizontal Distance then = K $\times$ R $\times$ cos$^2\alpha$, and the Difference in elevation then = K $\times$ R $\times$ cos $\alpha \times$ sin α. Sometimes the DE is given as ½ KR sin 2α. This is a trigonometric identity, i.e., ½ sin 2α = cos α sin α.

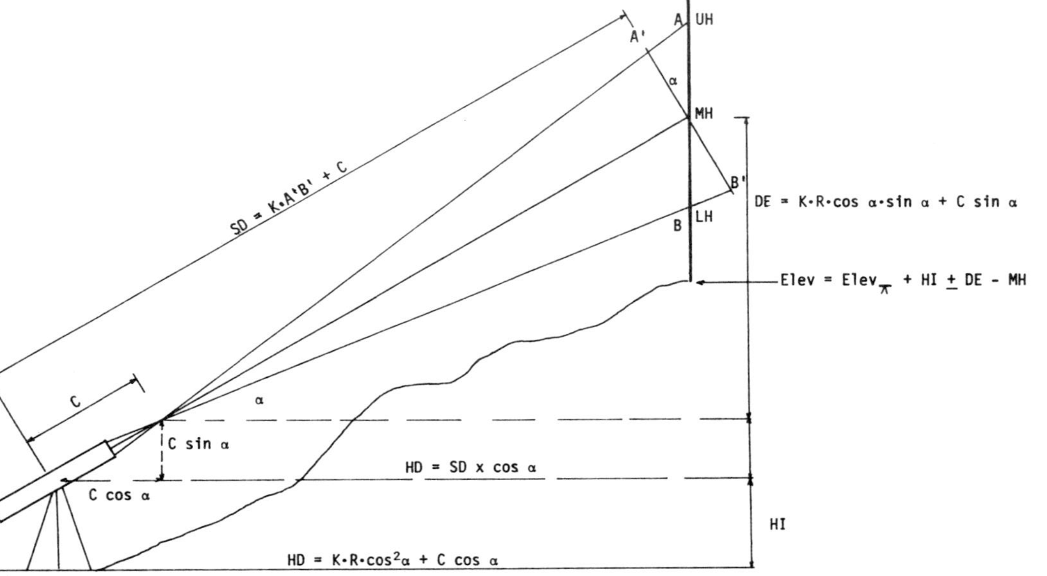

Figure 6-3. Inclined sight using stadia method.

The horizontal distance thus computed is the distance between the focal point and the rod if $C \neq 0$ and the difference in elevation is that between the focal point and the intersection of the line of sight of the middle stadia hair on the level rod. If the instrument has a C value = 0, the horizontal distance and the difference in elevation are both measured from the horizontal axis.

The reduction of the stadia data to horizontal distance and difference in elevation is easily done with the aid of the tables found in the appendix. The table may be used with either vertical angles or zenith angles, remembering that vertical angles are measured from the horizontal plane and zenith angles are measured from a point directly overhead. The table gives values in increments of 02' for all vertical angles from $0°$ to $30°$. The angles found at the top of each column use the minutes that read from the top down and the zenith angles located at the bottom of each column use the minutes in the column that reads up. All zenith angles can be converted to vertical angles. For purposes of further explanation, only vertical angles will be used since $90°$ – zenith angle = the vertical angle. The value obtained from the table under the heading of HD is really $100 \cdot \cos^2\alpha$ and the value obtained under the heading of DE = $100 \cdot \cos \alpha \cdot \sin \alpha$. These values, found in the table, when multiplied by R given the horizontal distance and difference in elevation.

At the bottom of each column are found C values of 0.75, 1.00, and 1.25 and under the HD and DE columns are values which equal $C \times \cos$ and $C \times \sin$ respectively for the particular vertical angle.

In stadia work it is usually the ground elevation of the point on which the rod is placed that is needed. The elevation of the point is easily found using the following equation.

Elevation = Elev. + HI ± DE − MH

Elev.　　　　= ground elevation of point on which instrument is set.

HI　　　　= vertical distance from ground to horizontal axis of the transit.

DE　　　　= difference in elevation (K × R cos α × sin α) may be + or −.

MH　　　　= sight of middle hairline on stadia rod.

It should be apparent that if the elevation of many points were needed, the MH should be sighted on the HI because they would then cancel each other in the equation for the elevation of a point.

There are several methods or procedures for obtaining field data in a stadia traverse.

1. Place the middle hair on the HI of the instrument and read the upper and lower stadia hair lines and the vertical angle. The disadvantage of this is that the rod intercept is not readily determined and that the HI point on the rod may be obscured.

2. Place the lower stadia hair on a foot mark and record the middle and the upper hairlines and the vertical angle.

The disadvantage is that the middle stadia hair is not sighted on the HI and more calculations are needed to solve for the elevation of the point.

3. Sight the middle hair on the HI and then raise or lower the line of sight until the bottom hair is on the nearest foot mark. The upper and lower stadia hairs are recorded and then the line of sight is changed moving the MH back to the HI after which the vertical angle is read and recorded. The disadvantage is that the MH is not midway between the recorded value of the UH and LH.

Tables are not always available to reduce stadia readings to horizontal distance and difference in elevation. Most transits and telescopic alidades have a device called a Beaman Stadia Arc on the vertical circle, scribed with scales which are used to compute horizontal distance and difference in elevation without having to read the vertical angle. The scales (Horz. and Vert.) are read from an index mark and have no vernier; the value read on either scale is equal to the value that would be found in the stadia reduction table under a corresponding vertical angle. The value $100 \times \cos^2\alpha$ for an inclined line of sight is read directly from the horizontal scale of Beaman Stadia Arc.

There are two methods of calibrating the "Vert." scale of Beaman Stadia Arc. One type reads zero when the telescope is level. The calibrations on the "Vert." scale give the value of $100 \times \cos\alpha \times \sin\alpha$. The instrument man must be very careful to note whether the sight is plus or minus for very small angles. The second type of "Vert" stadia scale reads 50 when the telescope is level. The DE is computed

Fig. 6-4. Transit (+ vertical angle on vertical circle) photo by Warren Welch).

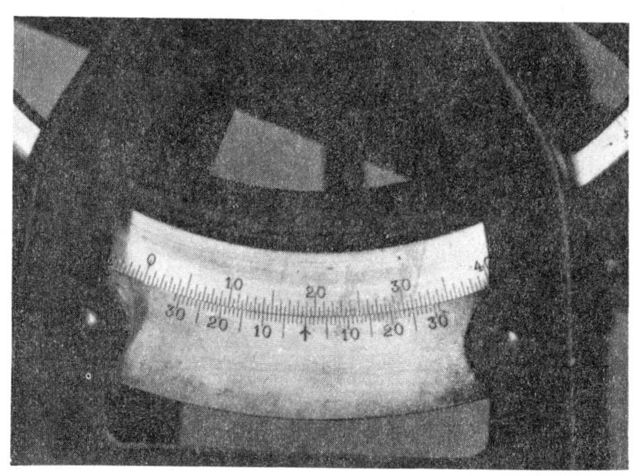

Fig. 6-5. Vertical scale and vernier (photos by Warren Welch).

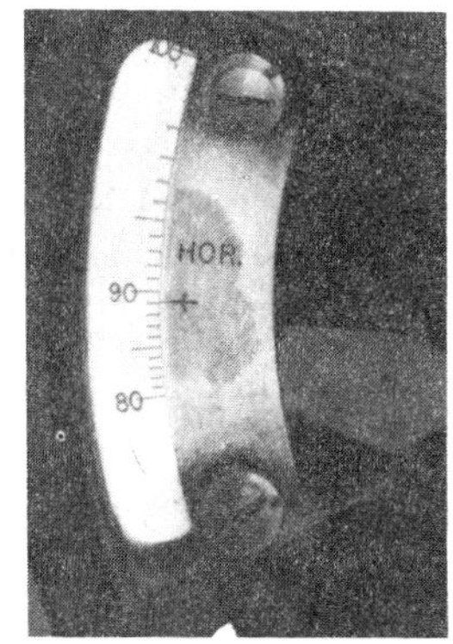

Fig. 6-6. Beaman "Horz" scale.

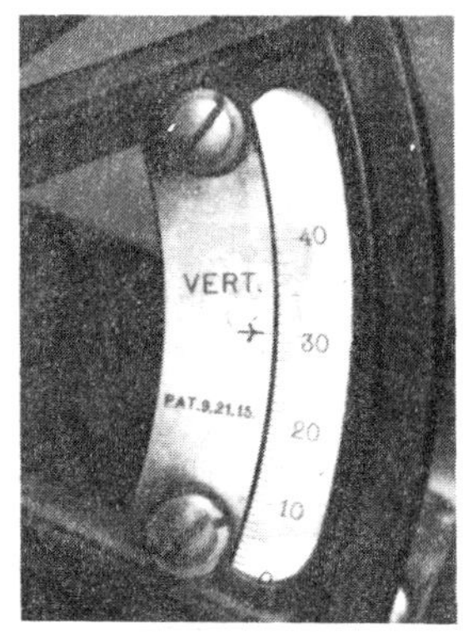

Fig. 6-7. Beaman "Vert" scale.

by multiplying the rod intercept (R) by the scale reading
– 50. With this type of scale, the vertical angle is + when
the vertical scale reading is greater than 50 and the difference in elevation will likewise be positive, and the difference in elevation will be negative when the scale reading is
less than 50.

Figure 6-4 is a picture of a transit in the position of
reading a positive vertical angle. The value of the vertical
angle as seen in an enlargement of the vertical scale and
vernier shown in Figure 6-5 is a + 18°53′. The reduction
factor of 89.525 which = (100 × cos 18°53′ × cos 18°53′)
as obtained from the table is almost identical to the value
of 89.1 read on the horizontal scale of the Beaman Stadia
Arc shown in the enlargement in Figure 6-6. Figure 6-7, an
enlargement of the "Vert" scale with a reading of approximately 31 compares favorably with the value of 30.63
(100 × cos 18°53′ × sin 18°53′) obtained from the table.
The slight difference may be due to the fact that the angle
used for this example is quite large.

The problem of having to consider the C value with an
external focusing transit may sometimes be eliminated by
increasing the observed R value by 0.01 when the C value
= 1 as shown in the following example.

Given: HI = 4.90′
          Vertical angle = +6°12′
          UH = 6.08
          MH = 5.04   R = 2.08
          LH = 4.00
          Elevation of transit station = 695.00′
          C = 1.00

Method 1.

    Horz. Dist. $= K \times R \times \cos^2\alpha + C \cos \alpha$

               $= (2.08)(98.83) + 0.99$

               $= 206.56'$

    Diff. Elev. $= K \times R \times \cos \alpha \times \sin \alpha + C \sin \alpha$

               $= (2.08)(10.74) + 0.11$

               $= +22.45'$

    Elevation $= \text{Elev} + \text{HI} + \text{DE} - \text{MH}$

               $= 695.00 + 4.90' + 22.45' - 5.04'$

               $= 717.31'$

Method 2. (C value added to R = 2.08 + 0.01)

    Horz. Dist. $= (2.09)(98.83)$

               $= 206.55'$

    Diff. Elev. $= (2.09)(10.74)$

               $= +22.45'$

If the vertical angle, the rod intercept and the reading on the middle stadia hair are known the horizontal distance and the elevation can always be computed. It does not make any difference where the sighting is taken on the level rod as long as the necessary data can be obtained. A quick check, as data is taken, is to make certain that the interval between the lower and middle stadia hairs does not differ from the interval between the middle and upper hairs by more than 0.01 of a foot.

1.   Using the stadia method, the horizontal distance is computed to be **294.08** feet when the rod intercept R (UH-LH) = 3.08 feet.   Find the slope distance between the two points and the zenith angle.

2. Sta / Azimuth / Vert angle / LH / MH / UH

Instrument at A Elev. = 963.5 feet   HI = 4.95'   C=1.25

   B   / 187°43'   / + 19°38'   / 8.22/ 9.28 / 10.33

Find the slope in % between A and B.

3.   Instrument at station D. Backsight on True North Azimuth = 0 00' HI = 4.98' C = 1.00

Station	Azimuth	Zenith Angle	LH	MH	UH
E	211°42'	97°07'	8.00	9.65	11.31
F	347°16'	78°23'	2.00	2.87	3.74

Find the azimuth of EF and the slope from E to F in percent.

LEVELING

Leveling is the process of determining the elevation of points or the difference in elevation between the points. The elevation of a point is the vertical distance above or below a reference datum. The reference datum is usually taken to be mean sea level. A horizontal line is tangent to a level surface and a level line is a line in a level surface. A vertical angle is the angle between two intersecting lines in a vertical plane, one of which lines is horizontal.

There are three methods of determining differences in elevation.

1. Barometric leveling is measuring differences in barometric pressure to determine elevations of various points.

2. Trigonometric leveling is mathematically computing differences in elevation using vertical angles and slope or horizontal distances which have been measured.

3. Direct or spirit leveling measures vertical distances directly and of the three is the most accurate method.

Only in the most precise type of leveling is it necessary to consider the effects of curvature and refraction. Curvature is the vertical distance between a level line and a horizontal line. Refraction is the change in direction of light rays as they encounter mediums of varying density, in this case the earth's atmosphere. Light waves near the surface of the earth are bent downward. The combined effect of curvature and refraction is expressed as 0.66 feet (curvature/mile) - 0.09 feet (refraction/mile) or $0.57K^2$ where

K is the distance in miles. The combined effects of curvature and refraction is also given as $0.0206'M^2$ or $0.021'M^2$ where M is the distance in thousands of feet. Thus in a sight of 200 feet (distance from level to level rod) would require a correction of $0.021'(200/1000)^2$ or $0.0008'$ which is too small to consider.

Level rods come in a variety of types with different graduations. They may be one piece rods or made in sections which are extendable to 12 or 13 feet. Two general classes are (1) self reading rods which are read by the instrument man or (2) target rods on which a target is attached to the rod and the rod is lengthened or shortened as directed by the instrument man but the readings are made by the rodman. The most common type of self reading rod is the Philadelphia rod, shown in Figure 7-1. It is graduated in 1/10ths and 1/100ths of feet. The background is white and the graduations, which are 0.01 feet wide, are in black. The big black numbers indicate tenths of feet, and the big red numbers indicate feet. This rod is also considered to be a target rod because the graduations on the back side of the rod shown in Figure 7-2 go in a reverse direction, i.e., from the top down. There is also a vernier which enables the rodman to read the height of the target above the bottom of the rod after he has adjusted the target by extending or shortening the rod.

Two types of engineer's levels that are generally used in surveying are (1) Dumpy Level Figure 7-3 and (2) Self-Leveling Level Figure 7-4. Figure 7-5 and Figure 7-6 show a Kern GKO-A level which has no leveling screws. The conical bearing surface of the instrument base sits on the spherical surface of the tripod head shown in Figure 7-7. Coarse leveling is accomplished by shifting the instrument on this spherical surface until the bull's-eye level is center-

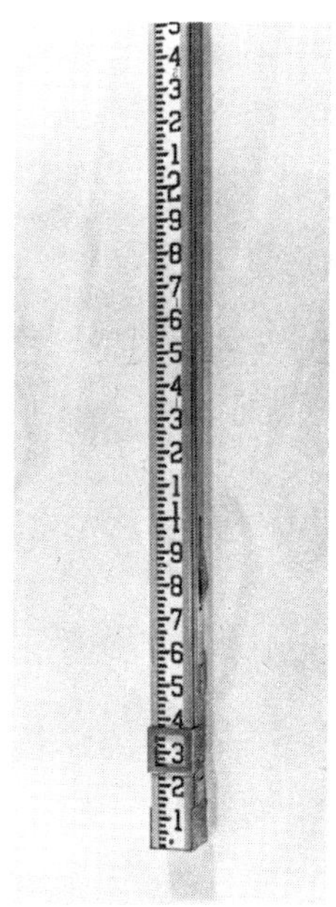

Figure 7-1. Front face of
direct reading rod
(courtesy Lietz Co.).

Figure 7-2. Back side of
target rod
(courtesy Lietz Co.).

Figure 7-3. Dumpy level (courtesy Keuffel & Esser Co.).

ed. When the coarse leveling is finished, the compensator automatically completes the fine leveling. The compensator unit has an optical image-reversion system. It is suspended with its counter weight from a horizontal axis. When the coarse leveling is inadequate, a red warning diaphragm appears at either the upper or lower edge of the field of view of the telescope. The compensator is shown in Figure 7-8.

Figure 7-4. Self-Leveling level (Courtesy Lietz Co.).

Figure 7-5. GKO-A Level
(Courtesy of Kern & Co., Ltd.)

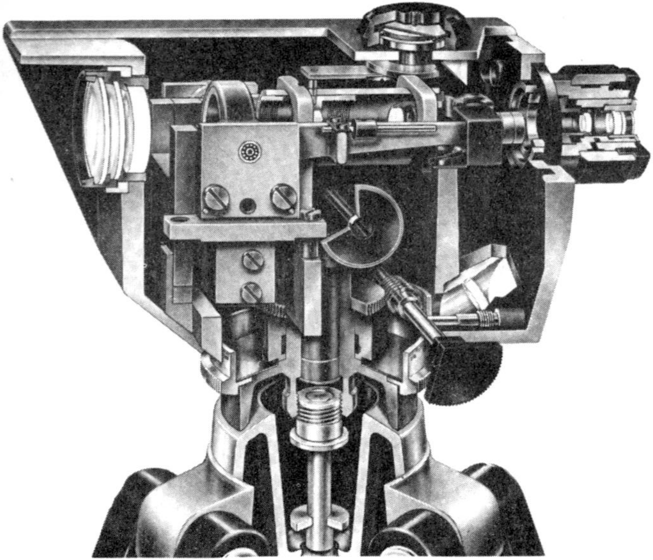

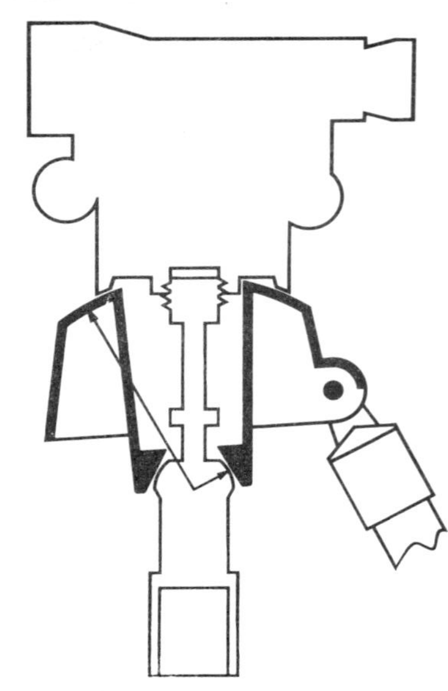

Figure 7-6. Cut away view of GKO-A (Courtesy of Kern & Co., Ltd.).

Figure 7-7. Cut away view of Tripod head of GKO-A level (Courtesy of Kern & Co., Ltd).

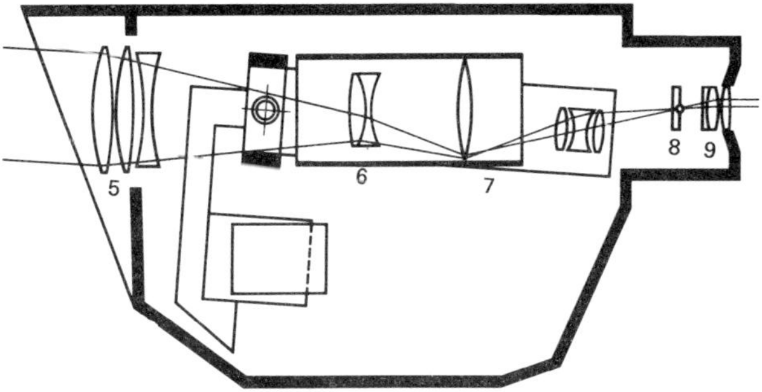

Figure 7-8. Cut away view of GKO-A Compensator
(Courtesy of Kern & Co., Ltd.).

Figure 7-9. Bullseye Level on Self-Leveling Level.

The terminology used in leveling is similar to that used in transit work, but some terms have completely different meanings.

B.M. A Bench Mark is a point whose elevation is known and which has a permanent and known location. These bench marks are established through the United States by the U.S. Geological Survey and the U.S. Coast and Geodetic Survey. They are usually bronze plates set in stone or concrete.

T.B.M. Temporary Bench Marks are set by engineers for construction purposes. They may be marks on curbs or nails in trees. They are not set permanently.

T.P. A Turning Point is any point on which the rod is placed for taking backsights and foresights. A T.B.M. or B.M. may be used as a turning point.

B.S. A Backsight is a rod reading taken on a rod when the rod is placed on a point of known elevation. It is sometimes called a plus (+) sight because it is always added to the known elevation of the point to obtain the H.I.

F.S. A Foresight is a rod reading taken on a rod when the rod is on a point whose elevation is unknown. Sometimes called a minus (–) sight because it is always subtracted from the H.I. to obtain the elevation.

H.I.   The Height of Instrument is the elevation of the line of sight when the instrument is level.

The two kinds of leveling projects that are most common are: (1) differential leveling and (2) profile leveling.

(1)   Differential leveling is the process of determining the elevation of points some distance apart by carrying the elevation from one point to another.

(2) Profile leveling is the determination of the elevation of points at predetermined intervals along an established line such as the center-line of a proposed road. With data obtained, the profile of the ground line may be plotted, and then grade lines of finished projects may be planned.

Differential leveling notes are simple and easy to keep if the following facts are always kept in mind.

$$\text{Elev.} + \text{BS} = \text{HI} \quad \text{and} \quad \text{HI} - \text{FS} = \text{Elev.}$$

The notes and note form for the data shown in Figure 7-8 would be as follows.

Station	B.S.	H.I.	F.S.	Elev.
TBM	6.22	306.22	---	300.00
T.P.	5.18	311.27	0.13	306.09
T.P.			2.72	308.55

There is never a foresight taken on the first station, only a backsight.

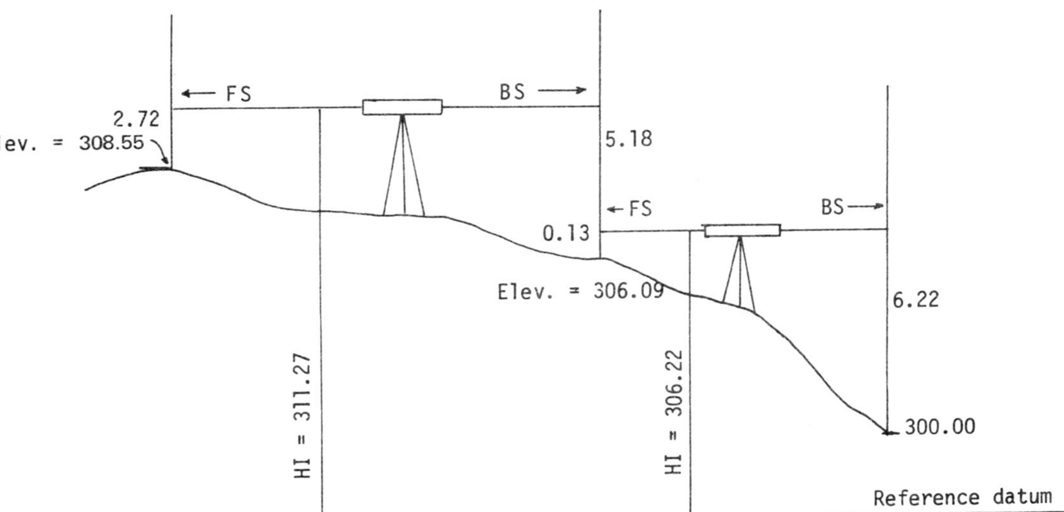

Figure 7-10. Differential leveling sketch.

The level is first set up at a convenient height for viewing and roughly leveled. The level is not set up over a definite point, it may be anywhere from which the backsight station can be conveniently seen, and so located that foresights can be taken in the direction of travel of the leveling project. Only leveling of the instrument needs to be done, there is no centering over a point required. Use the legs to roughly level the foot plate when using the Dumpy or a Wye level, completing the fine adjustment centering the bubble level when oriented over a pair of opposite foot screws. Turn it 90° to orient the level over the remaining pair of foot screws and center the level. The bubble should remain centered as the level is turned through 360°. However, check the bubble before taking a sight and re-level or re-center the bubble each time. A self leveling level has a spherical ball joint (Figure 7-9) instead of a foot plate and a bull's eye level (Figure 7-10) instead of a bubble level, and three foot screws instead of four. Roughly level the instrument using the spherical level and bull's eye level, finish the leveling by turning the head to put the bull's eye level in line with two of the foot screws, then use one of the screws at a time to center the bubble.

It is important to make certain that the rod and hairlines are in clear focus. If the hairlines appear to move with a slight shift of the eye, parallax exists and must be eliminated by further focusing of the objective lens and the eye piece. The focusing may easily be done using the rod set on the beginning, or first backsight station.

The rod, which must be in a vertical position at the time of reading, may easily be lined up in the vertical plane by means of the vertical hair line. The instrument man can indicate to the rodman by arm signals which direction the rod must be leaned to line up with the vertical hairline. It is impossible to determine, when sighting through the level, whether the rod might be leaning toward or away from the instrument. This possible error is eliminated by having the rodman lean the rod slowly toward and then away from the level while holding the rod lightly between his fingers and keeping it in the plane parallel to the vertical hairline. During this procedure the horizontal hairline will appear to move up the rod and then down to a minimum reading and go up the rod again. The minimum rod reading is the one sought because at this point the rod is perpendicular to the line of sight. This procedure, shown in Figure 7-11 will have to be repeated several times until the instrument man gets the reading in feet, tenths and hundredths. This method of obtaining the minimum reading is sometimes difficult on windy days with the rod fully extended. Sometimes a bull's eye bubble is attached to the level rod and when the bubble is centered, the rod is in a true vertical position.

After the backsight has been taken, the rod is moved ahead to a convenient point which becomes a turning point (T.P.). The level is then sighted on the rod on the T.P. and a rod reading is taken and recorded on the line of the T.P. under the foresight column. The rodman must be careful to note the location of the T.P. because he has to rotate the rod $180°$ before a B.S. is taken and the rod has to be on the same point. Objects that are hard and firmly in place are usually chosen for turning points. The rod should never be placed on grass or sod when doing differential leveling.

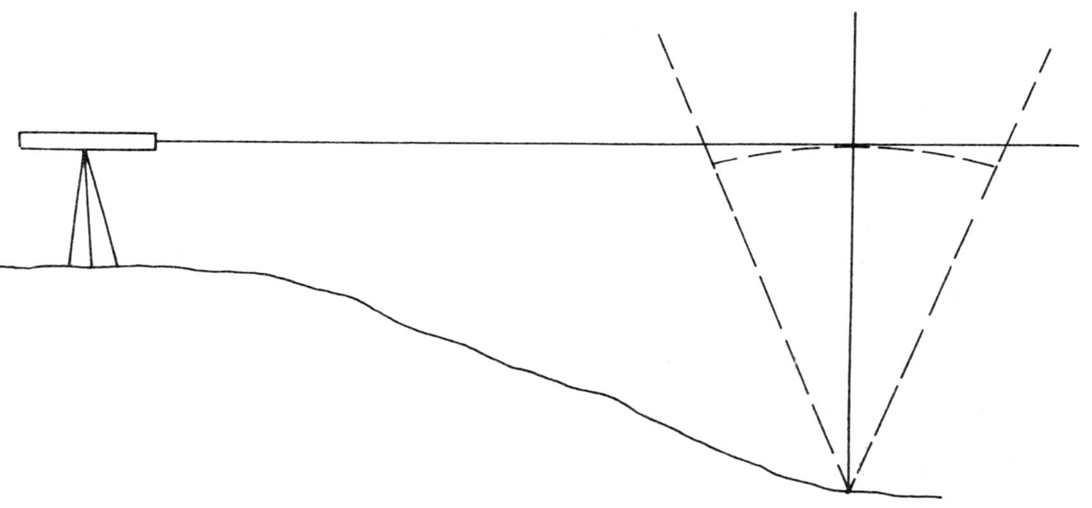

Figure 7-11. Rod reading perpendicular to line of sight.

The level is now moved ahead of the rod a distance that is approximately equal to the distance between the level and level rod when the foresight was taken. Roughly equalizing this distance eliminates the necessity of making any corrections for curvature and refraction. A backsight is taken on the rod, the rod is moved ahead, etc., until the project is completed.

A simple check can be made on the level notes by adding all of the B.S. and all of the F.S. The sum of the B.S. (+) and F.S. (−) should equal the difference in elevation between the first and last station. This does not account for an error made in reading the rod and/or recording incorrect values in the field book. If a level circuit (closed) is run, the error in closure may be adjusted based on the length of the lines or on the number of set ups. The chance of error increases with the length and number of set ups.

Profile leveling is the determination of the elevation of established points along the centerline of a proposed road location. The interval may be 50 or 100 foot stations or irregularly stationed points. The method is almost identical to that followed in differential leveling except that the instrument is not necessarily moved after every sighting on the rod on a point of unknown elevation, and the note form has a column heading I.F.S. (intermediate foresight)

STA   BS   HI   IFS   FS   ELEV

The project is started from a BM of known elevation or from the first station with an assumed elevation. The Elev. + the BS gives the HI of the instrument and this value does not change as long as the instrument is not moved. Any number of rod readings may be taken from one set up and they are called intermediate foresights. Only when the instrument is moved to another location is a BS taken on a T.P. In general, the IFS's are read only to the nearest tenth of a foot, while the FS is read to the nearest hundredth. The BS and FS are treated in the same manner as they are in differential leveling. The IFS's are a form of a FS since the elevations of the points are to be determined, but they are recorded in a different column. The elevation of the ground at each rod setting is obtained by subtracting the IFS from the HI, which does not change at any one set up.

The data thus gathered in profile leveling is used to plot the ground profile and to aid in engineering the grade of the proposed road. It is common practice to plot the horizontal distances and elevations to two different scales. For example; the Horizontal scale might be 1″ = 100′ while on the vertical scale 1″ = 10′. Once the ground profile is plotted, trial road grades may be tested until a suitable one is found.

1.   The amount of correction that must be made for curvature and refraction equals 4.637 feet.   Find the length of this sight in feet (HD) between instrument and level rod.

2. Fill in all blanks in the note form below.

Station	BS	HI	IFS	FS	Elev.
0+00	______	886.53			875.00
0+50			9.6		______
1+00			______		879.2
1+50	10.62	______			883.46
2+00			______		886.0
2+50			5.4		______
3+00	______	903.61		1.84	______
3+50				______	899.04

3. A differential leveling project is carried on through a tunnel with the following data being taken

BS rod reading = 9.25 Rod placed on B.M. whose elevation = 1200 feet

FS rod reading = 12.06 Rod inverted with zero end of rod placed on TP_1 which is on roof of tunnel.

BS rod reading = 5.64 Rod still inverted and held on TP_1

FS rod reading = 3.88 Rod placed on TBM_1 in normal position

Find the elevation of TBM_1.

4. Fill in all blanks in set of leveling notes below.

Station	BS	HI	FS	Elev.
BM_1	12.04	______		816.43
TP_1	______	833.61	4.37	______
TP_2	6.31	______	2.54	______
TP_3	3.68	______	______	825.89
TP_4	______	839.63	1.05	______
BM_2			2.83	______

Find the horizontal distance between the two bench marks if the slope between them equals +3.5%.

PLANE TABLE

The plane table method of mapping is advantageous because when the map is completed in the field no further calculations are needed. A disadvantage is that no data is collected, thus no check can be made if an error occurs in the map. The term "plane table" usually implies a tripod attached to and supporting a drawing board on which the map is constructed with the aid of some type of alidade. Plane table boards come in a variety of sizes but are usually 15" by 24" or 18" by 24". The board is attached to the tripod by means of a "Johnson" head, which is a form of ball and socket joint. The "Johnson" head permits the board to be leveled and locked in the level position before orienting the board in direction. After the board is leveled the board is rotated on a vertical axis and thus oriented.

With the board is used an alidade, which is a combination sighting device and straight edge, or ruler. A peep sight alidade consists of two sighting vanes, similar to those of a staff compass, attached to the ends of a brass ruler, one side of which is scribed with a suitable scale for mapping, the vanes being so placed that a sight through the vanes is parallel to the ruler edge.

To use the plane table a piece of drawing paper is attached to the board and the board and tripod are set up in the field in a position suitable for viewing the terrain to be mapped. The instrument should be of a convenient height for the mapper to work on, usually a little more than waist high. The board is leveled by eye and oriented either by use of a compass or merely by eye. Orientation is not of great importance because all points of the map are drawn in position relative to other points. A beginning point is chosen on the paper; this point should be placed on the paper in about the same relative position to the expected finished drawing as the plane table itself is to the terrain surrounding it. A pin is stuck in the paper at the chosen point, rigid enough that it will remain in place when the alidade is pressed against it. Call this point A.

When mapping using the peep sight alidade objects or points are located by the intersection of two lines of sight, which when plotted are called rays. The alidade is placed on the board and against the pin; using the pin as a center of rotation the sight is brought on the target and a line or ray is drawn from the pin in the direction of the line of sight, i.e., along the ruler. This is repeated for each object or point that is desired to be on the map, resulting in a number of lines being drawn in various directions from the pin, Point A, as a center. If very many lines are drawn it is necessary to keep track in some manner of what object they are drawn to. When all the necessary rays have been drawn in, the board must be moved to a new location, as follows. Select the new location, sight on it, and draw a ray toward it. Measure the distance between the board, Point A, and the new position, Point B. Measure this distance, to scale, along the ray toward the new position and label it B. Pick up the equipment and move to Point B and set up the board again. This time orient the board by placing the alidade along the ray from A to B, turn the board bodily on the vertical axis until the sight is on Point A; the board will now be oriented in the same direction as it had been at Point A. Now take sights on all the points previously sighted from Point A, drawing rays as before. The intersection of the rays drawn to the same point from the two points A and B will locate the point or object in question. In this manner all required points are located. Rays drawn to new points or objects from B can be crossed with rays from another traverse point C, where the table is again set up and oriented as explained above. Proceed until the total traverse is completed and all required points for the map you want are located.

A word of caution is in order here. A large number of rays on the paper will soon confuse the mapper so as soon as objects can be located on paper erase all the unnecessary rays. Trace the object (house, street, ect.) clearly and eliminate **everything** on your mapping paper that you possibly can. This simply cannot be overemphasized.

The other method of locating detail, which requires an angle and a distance, uses a telescopic alidade (Figure 8-1). The telescopic alidade is a metal ruler on which is mounted a telescope similar to that of a transit, complete with vertical scales and verniers, Beaman Stadia Arc, and stadia cross hairlines. The same procedure of setting up, leveling and orienting the table is followed. The blade of the alidade is placed against the pin and a sight is taken on the leveling rod placed on the object or point which is to be plotted on the map sheet. The horizontal distance is determined by the stadia method. It must be remembered that the stadia arc must be read and used if there is any appreciable vertical angle involved. One edge of the blade has a scale against which the distance is scaled off and the point is marked on the map sheet rather than drawing a pencil line. When two or more points such as corners of a building are located, they should immediately be connected so that there will be no guesswork as to what points represent what objects.

Figure 8-1. Planetable RK with reducing alidade
(Courtesy Kern & Co., Ltd.).

Before moving to a new station, be sure to plot the line and distance to the new station on the map sheet. The table is moved to the new station, oriented in the same manner described for the peepsight alidade and the process of radiation in locating detail is continued.

The foregoing discussion has dealt only with planemetric detail. Topographic detail is easily obtained but can be done only with the telescopic alidade. Two methods of obtaining contours are (1) by determining elevations of the breaks in the slope (point where the slope changes) and interpolation for points of desired elevation (2) plotting of known elevation. The latter is the best method if the area being mapped is not too extensive. The ground elevation of the plane table station needs to be known as well as the HI of the telescopic alidade. The alidade is leveled-and used like an engineer's level. The rod is moved around until it is finally placed on a point whose elevation is that of the contour desired and the horizontal distance is computed by the stadia method and plotted. In this manner as many points as are desired of the same elevation are located and the contour line is created when all of these points (same elevation) are connected. Traversing with a plane table is simple but it must be remembered that the accuracy is poor.

APPENDIX

Flow Chart — steps in doubling field angles

Tape Corrections

Instrument Adjustments

Trigonometric Formulae

Decimal Degrees

Stadia Reduction Table

Slope Reduction Table

Trigonometric Tables

MEASUREMENT OF HORIZONTAL AND VERTICAL ANGLES

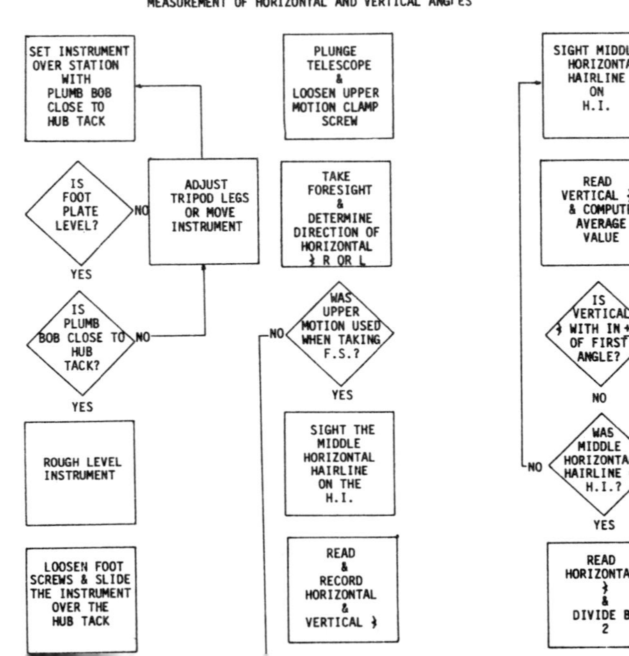

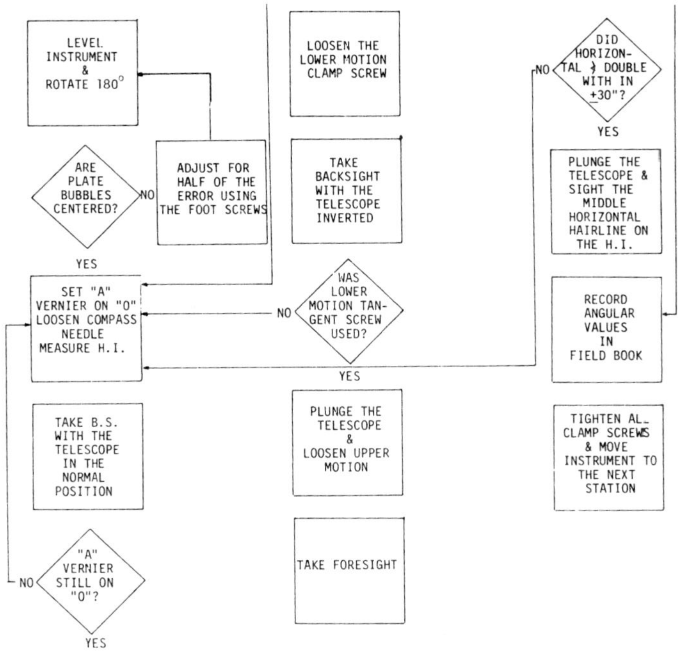
LEVEL INSTRUMENT & ROTATE 180°
ARE PLATE BUBBLES CENTERED?
YES
NO
ADJUST FOR HALF OF THE ERROR USING THE FOOT SCREWS
LOOSEN THE LOWER MOTION CLAMP SCREW
TAKE BACKSIGHT WITH THE TELESCOPE INVERTED
DID HORIZON-TAL } DOUBLE WITH IN +30"?
NO
YES
PLUNGE THE TELESCOPE & SIGHT THE MIDDLE HORIZONTAL HAIRLINE ON THE H.I.
SET "A" VERNIER ON "0" LOOSEN COMPASS NEEDLE MEASURE H.I.
WAS LOWER MOTION TAN-GENT SCREW USED?
NO
YES
RECORD ANGULAR VALUES IN FIELD BOOK
TAKE B.S. WITH THE TELESCOPE IN THE NORMAL POSITION
PLUNGE THE TELESCOPE & LOOSEN UPPER MOTION
TIGHTEN ALL CLAMP SCREWS & MOVE INSTRUMENT TO THE NEXT STATION
"A" VERNIER STILL ON "0"?
NO
YES
TAKE FORESIGHT

Tape Corrections

Corrections are easily applied if the tape is not a standard length by the following equation.

$$L = \frac{T}{S} M$$

L = True length of line
T = True length of tape
S = Standard length of tape
M = Measured (recorded) distance

The T value can only be determined by comparing the tape to a standard distance or by buying a tape which has been standardized. (Length determined for a given tension and temperature.)

If the tape is supported only at the ends (most tapes are supported throughout their entire length when standardized) it will then hang in the form of a catenary curve. A correction may be applied for what is termed sag by the following formula.

$$C_s = \frac{w^2 L^3}{24t^2}$$

C_s = Correction in feet
w = Weight of tape in lbs/foot
L = Distance between supports in feet
t = Tension in pounds

This correction will always be negative because the true distance will always be less than the measured distance. This correction is also based on the assumption that both ends of the unsupported tape are at the same elevation and never occurs in rough terrain. A more nearly correct value is obtained by computer solution which is used in skyline design for logging and is described in Research Paper 35,

December 1976 Published by the Forest Research Laboratory, School of Forestry, Oregon State University.

Correction made for tensions other than that at which the tape was standardized can be computed as follows.

$$C_p = \frac{L(t-t_0)}{SE}$$

C_p = Correction per tape length in feet

L = Length of tape in feet

t = Tension applied in pounds

t_0 = Tension (standard) in pounds

S = Cross sectional area of tape in sq. in.

E = Modulus of elasticity (29,000,000 #/sq. in).

This correction can be applied only when the tape is supported throughout its entire length and this condition is seldom encountered in the field.

Correction for temperature is only made when very precise measurements might be needed such as in the base line for triangulation, and even these measurements can be more accurately obtained with electronic equipment. The thermal coefficient of expansion is 0.00000645 feet per 1 degree Fahrenheit which means that the change in the length of a 100 foot tape will be 0.0097 feet or approximately 0.01 feet for every 15 degrees Fahrenheit change in temperature. The following formula is used for temperature corrections.

$$C_t = K(T_t - T)L$$

C_t = Correction in length due to temperature change

K = Thermal coefficient of expansion of steel (0.00000645)

T_t = Temperature at time of measurement

T = Temperature at standard length

L = Measured (recorded) length of line

Adjustment of Transit

1. Make vertical cross hair lie in plane perpendicular to horizontal axis: Sight vertical hairline on a well defined point 200 feet away and rotate telescope through a small vertical angle. If the point appears to depart from the vertical hair, loosen two adjacent capstan screws and rotate the reticle ring until the vertical hairline traverses on the point through the entire length of the hairline, and then tighten the screws.

2. Make the axis of each plate level be in a plane perpendicular to the vertical axis: Rotate the instrument about the vertical axis until each level vial is parallel to a plane passing through opposite leveling screws and center bubbles in the level tube. Rotate the instrument 180° about the vertical axis. If the bubbles are displaced bring them back half of the displacement by means of the adjustment screws. There should be no displacement of the bubble as the instrument is rotated 360°.

3. Make the line of sight perpendicular to the horizontal axis: Level the instrument and sight on a point A at least 300 feet distant with the telescope in the normal position with both motions clamped. Plunge the telescope and set point B on the line of sight at about the same distance on the other side of the transit. Loosen the upper motion and sight on point A again with the telescope inverted. Clamp the upper motion, plunge the telescope again and set point C beside B if line of sight does not fall on B. Mark point D one quarter of the distance from C to B and adjust reticle ring by using opposite screws until the line of sight passes through D. The distance between C and B can easily be determined if the sights are made on a level rod placed horizontally on the ground approximately perpendicular to line of sight to B & C.

4. Make horizontal axis perpendicular to the vertical axis: Level the transit very carefully and with the telescope in the normal position, sight on a point A at some considerable height above the instrument. With the horizontal motions clamped, depress the telescope and set B on the ground. Rotate the instrument about the vertical axis and with the telescope in the inverted position sight on point A. Depress the telescope and if the line of sight does not intersect point B, set C alongside B. Establish point D halfway between B and C. Sight on D and then raise the telescope until line of sight is beside A. Bring the line of sight onto A by loosening the screws of the bearing cap and then raising or lowering the adjustable end of the horizontal axis.

5. Make the axis of the telescope level parallel to the line of sight: Set two pegs A & B approximately 200-300 feet apart. Set the instrument with the eyepiece very close to a level rod held on point A and take a rod reading (a) by sighting through the objective lens. Move the rod to point B and take rod reading (b) with the transit still near A. Move the instrument to B, set up as before and take rod readings (c)and (d) and B & A respectively. If the two differences in elevation (a–b) and (d–c) are equal the instrument is in adjustment. If not, the correct rod reading d' with the rod on A and instrument remaining at B is equal to C + ((a–b)+(d–c))/2. This equation is solved with due regard to signs. The adjustment is made by raising or lowering one end of the telescope level tube until the bubble is centered while the sight is on d'.

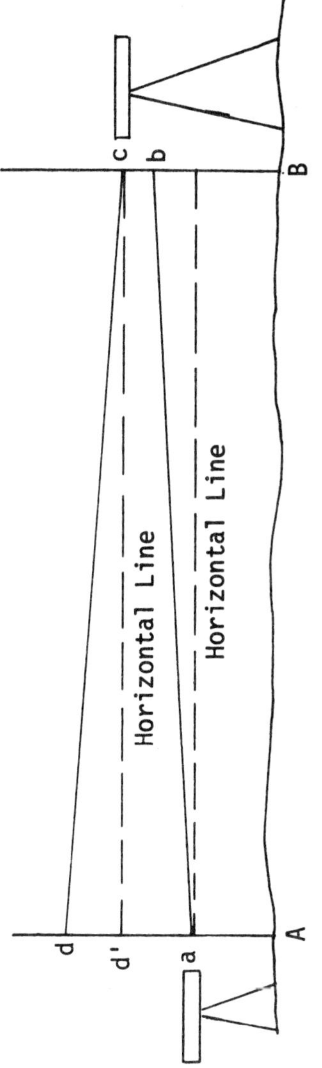

A
c
b
B
Horizontal Line
Horizontal Line
d
d'
a
A
B

Adjustment of Dumpy Level

1. Make the axis of the level tube perpendicular to the vertical axis: Center the bubble approximately over each pair of opposite leveling screws and then center carefully over one pair. Rotate the level 180° about its vertical axis. If the bubble is displaced, bring it back half way by means of the capstan nuts at one end of the level tube. Re-level the instrument using the leveling screws and repeat the adjustment until there is no displacement of the bubble.

2. Make the horizontal cross hair lie in a plane perpendicular to the vertical axis: Sight the horizontal cross-hair on a well defined point and rotate the instrument slowly about its vertical axis. If the point appears to depart from the cross-hair, loosen two adjacent capstan screws holding the reticle ring and rotate the reticle ring which holds the cross-hair until by further trial, the point appears to travel the entire length of the cross-hair, and then tighten the screws which were loosened.

3. Make the line of sight parallel to the axis of the level tube: Set two points A & B, 200-300 feet apart on approximately level ground. Set the level up at station A so that the eyepiece is less than 1″ from the level rod held on this station and take a rod reading (a) by sighting through the objective end of the telescope. Move the rod to station B and take rod reading (b) with the level at A.

Move the level to Station B and repeat the procedure taking rod readings (c) and (d) on stations B & A respectively. If the two differences in elevation (a-b) and (d-c) are equal, the line of sight is in adjustment. If not, the correct rod reading at A with the instrument at B is $d' = c + ((a-b)+(d-c))/2$. This equation must be solved with due regard to signs. The line of sight is adjusted and brought onto d' by moving the cross hair ring vertically. The process is repeated as a check.

Trigonometric Formulas

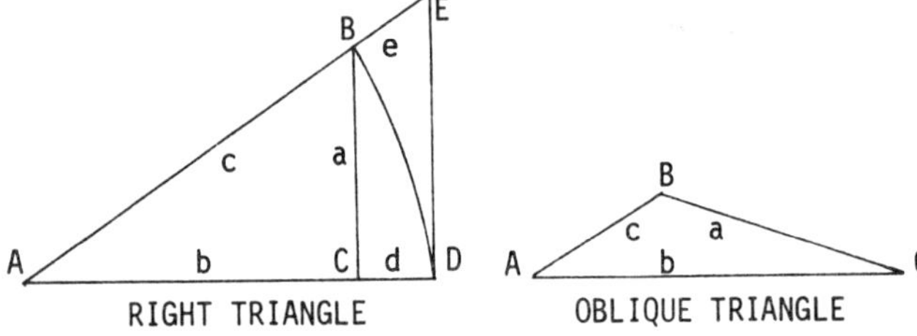

Solution of Right Triangles

$$\sin A = \frac{a}{c} = \cos B \qquad \cos A = \frac{b}{c} = \sin B \qquad \text{vers } A = \frac{c-b}{c} = \frac{d}{c}$$

$$\tan A = \frac{a}{b} = \cot B \qquad \cot A = \frac{b}{a} = \tan B$$

$$\sec A = \frac{c}{b} = \text{cosec } B \qquad \text{cosec } A = \frac{c}{a} = \sec B \qquad \text{exsec } A = \frac{e}{c}$$

$$a = c \sin A = b \tan A = c \cos B = b \cot B = \sqrt{(c+b)(c-b)}$$

$$b = c \cos A = a \cot A = c \sin B = a \tan = B = \sqrt{(c+a)(c-a)}$$
$$= c - c \text{ vers } A$$

$$d = c \text{ vers } A \qquad\qquad e = c \text{ exsec } A$$

$$c = \frac{a}{\cos B} = \frac{b}{\sin B} = \frac{a}{\sin A} = \frac{b}{\cos A} = \frac{d}{\text{vers } A} = \frac{e}{\text{exsec } A}$$
$$= b + b \text{ exsec } A$$

Solution of Oblique Triangles

Given	Sought	Formulas
A,B,a	b,c	$b = \dfrac{a}{\sin A} \cdot \sin B, \qquad c = \dfrac{a}{\sin A} \sin(A+B)$
A,a,b	B,c	$\sin B = \dfrac{\sin A}{a} \cdot b, \qquad c = \dfrac{a}{\sin A} \cdot \sin C$
C,a,b	A–B	$\tan \tfrac{1}{2}(A-B) = \dfrac{a-b}{a+b} \tan \tfrac{1}{2}(A+B)$
	c	$c = \sqrt{a^2+b^2-2ab \cos C}$
a,b,c	A	$\cos A = (b^2+c^2-a^2) \div 2bc$
		If $s = \tfrac{1}{2}(a+b+c)$, $\sin \tfrac{1}{2} A = \sqrt{\dfrac{(s-b)(s-c)}{bc}}$
		$\cos \tfrac{1}{2} A = \sqrt{\dfrac{s(s-a)}{bc}}$;
		$\tan \tfrac{1}{2} A = \sqrt{\dfrac{(s-b)(s-c)}{s(s-a)}}$
		$\sin A = \dfrac{2\sqrt{s(s-a)(s-b)(s-c)}}{bc}$;
		$\text{vers } A = \dfrac{2(s-b)(s-c)}{bc}$
	area	$\text{area} = \sqrt{s(s-a)(s-b)(s-c)}$
A,B,C,a	area	$\text{area} = \dfrac{a^2 \sin B \cdot \sin C}{2 \sin A}$
C,a,b	area	$\text{area} = \tfrac{1}{2} ab \sin C$

General Trigonometric Formulas

$$\sin A = 2 \sin \tfrac{1}{2} A \cos \tfrac{1}{2} a = \sqrt{1-\cos^2 A} = \tan A \cos A$$
$$= \sqrt{\tfrac{1}{2}(1-\cos 2A)}$$

$$\cos A = 2 \cos^2 \tfrac{1}{2} A - 1 = 1 - 2 \sin^2 \tfrac{1}{2} A = \cos^2 \tfrac{1}{2} A - \sin^2 \tfrac{1}{2} A$$
$$= 1 - \text{vers } A$$

$$\tan A = \frac{\sin A}{\cos A} = \frac{\sqrt{1-\cos^2 A}}{\cos A} = \frac{\sin 2A}{1+\cos 2A}$$

$$\cot A = \frac{\cos A}{\sin A} = \frac{\sin 2A}{1-\cos 2A} = \frac{\sin 2A}{\text{vers } 2A}$$

$$\text{vers } A = 1 - \cos A = \sin A \tan \tfrac{1}{2} A = 2 \sin^2 \tfrac{1}{2} A$$

$$\text{exsec } A = \sec A - 1 = \tan A \tan \tfrac{1}{2} A = \frac{\text{vers } A}{\cos A}$$

$$\sin 2A = 2 \sin A \cos A$$

$$\cos 2A = 2 \cos^2 A - 1 = \cos^2 A - \sin^2 A = 1 - 2 \sin^2 A$$

$$\sin^2 A + \cos^2 A = 1$$

Geometric Formulas

Required	Given	Formulas
Area of		
Circle	Radius = r	πr^2
Sector of Circle	Radius = r, Arc = L_c	$rL_c/2$
Segment of Circle	Chord = C, Middle Ordinate = M	(2/3)CM (approximate)
Ellipse	Semi-axes = a and b	πab
Surface of		
Cone	Radius of Base = r; Slant Height = s	πrs
Cylinder	Radius = r, Height = h	$2\pi rh$ (surface not including the ends)
Sphere	Radius = r	$4\pi r^2$
Zone	Radius of Sphere = r, Height of Zone = h	$2\pi rh$
Volume of		
Prism or Cylinder	Area of Base = b; Height = h	bh
Pyramid or Cone	Area of Base = b; Height = h	bh/3
Frustum of Pyramid or Cone	Area of bases = b and b'; Height = h	$(h/3)(b+b'+\sqrt{bb'})$
Sphere	Radius = r	$(4/3)\pi r^3$

DECIMAL DEGREE EQUIVALENTS
OF MINUTES AND SECONDS

SEC	\(\) MINUTES					
	0	1	2	3	4	5
0	0	.01667	.03333	.05000	.06667	.08333
1	.00028	.01694	.03361	.05028	.06694	.08361
2	.00056	.01722	.03389	.05056	.06722	.08389
3	.00083	.01750	.03417	.05083	.06750	.08417
4	.00111	.01778	.03444	.05111	.06778	.08444
5	.00139	.01806	.03472	.05139	.06806	.08472
6	.00167	.01833	.03500	.05167	.06833	.08500
7	.00194	.01861	.03528	.05194	.06861	.08528
8	.00222	.01889	.03556	.05222	.06889	.08556
9	.00250	.01917	.03583	.05250	.06917	.08583
10	.00278	.01944	.03611	.05278	.06944	.08611
11	.00306	.01972	.03639	.05306	.06972	.08639
12	.00333	.02000	.03667	.05333	.07000	.08667
13	.00361	.02028	.03694	.05361	.07028	.08694
14	.00389	.02056	.03722	.05389	.07056	.08722
15	.00417	.02083	.03750	.05417	.07083	.08750
16	.00444	.02111	.03778	.05444	.07111	.08778
17	.00472	.02139	.03806	.05472	.07139	.08806
18	.00500	.02167	.03833	.05500	.07167	.08833
19	.00528	.02194	.03861	.05528	.07194	.08861
20	.00556	.02222	.03889	.05556	.07222	.08889
21	.00583	.02250	.03917	.05583	.07250	.08917
22	.00611	.02278	.03944	.05611	.07278	.08944
23	.00639	.02306	.03972	.05639	.07306	.08972
24	.00667	.02333	.04000	.05667	.07333	.09000
25	.00694	.02361	.04028	.05694	.07361	.09028
26	.00722	.02389	.04056	.05722	.07389	.09056
27	.00750	.02417	.04083	.05750	.07417	.09083
28	.00778	.02444	.04111	.05778	.07444	.09111
29	.00806	.02472	.04139	.05806	.07472	.09139
30	.00833	.02500	.04167	.05833	.07500	.09167
31	.00861	.02528	.04194	.05861	.07528	.09194
32	.00889	.02556	.04222	.05889	.07556	.09222
33	.00917	.02583	.04250	.05917	.07583	.09250
34	.00944	.02611	.04278	.05944	.07611	.09278
35	.00972	.02639	.04306	.05972	.07639	.09306
36	.01000	.02667	.04333	.06000	.07667	.09333
37	.01028	.02694	.04361	.06028	.07694	.09361
38	.01056	.02722	.04389	.06056	.07722	.09389
39	.01083	.02750	.04417	.06083	.07750	.09417
40	.01111	.02778	.04444	.06111	.07778	.09444
41	.01139	.02806	.04472	.06139	.07806	.09472
42	.01167	.02833	.04500	.06167	.07833	.09500
43	.01194	.02861	.04528	.06194	.07861	.09528
44	.01222	.02889	.04556	.06222	.07889	.09556
45	.01250	.02917	.04583	.06250	.07917	.09583
46	.01278	.02944	.04611	.06278	.07944	.09611
47	.01306	.02972	.04639	.06306	.07972	.09639
48	.01333	.03000	.04667	.06333	.08000	.09667
49	.01361	.03028	.04694	.06361	.08028	.09694
50	.01389	.03056	.04722	.06389	.08056	.09722
51	.01417	.03083	.04750	.06417	.08083	.09750
52	.01444	.03111	.04778	.06444	.08111	.09778
53	.01472	.03139	.04806	.06472	.08139	.09806
54	.01500	.03167	.04833	.06500	.08167	.09833
55	.01528	.03194	.04861	.06528	.08194	.09861
56	.01556	.03222	.04889	.06556	.08222	.09889
57	.01583	.03250	.04917	.06583	.08250	.09917
58	.01611	.03278	.04944	.06611	.08278	.09944
59	.01639	.03306	.04972	.06639	.08306	.09972
60	.01667	.03333	.05000	.06667	.08333	.10000

DECIMAL DEGREE EQUIVALENTS
OF MINUTES AND SECONDS

SEC	MINUTES					
	6	7	8	9	10	11
0	.10000	.11667	.13333	.15000	.16667	.18333
1	.10028	.11694	.13361	.15028	.16694	.18361
2	.10056	.11722	.13389	.15056	.16722	.18389
3	.10083	.11750	.13417	.15083	.16750	.18417
4	.10111	.11778	.13444	.15111	.16778	.18444
5	.10139	.11806	.13472	.15139	.16806	.18472
6	.10167	.11833	.13500	.15167	.16833	.18500
7	.10194	.11861	.13528	.15194	.16861	.18528
8	.10222	.11889	.13556	.15222	.16889	.18556
9	.10250	.11917	.13583	.15250	.16917	.18583
10	.10278	.11944	.13611	.15278	.16944	.18611
11	.10306	.11972	.13639	.15306	.16972	.18639
12	.10333	.12000	.13667	.15333	.17000	.18667
13	.10361	.12028	.13694	.15361	.17028	.18694
14	.10389	.12056	.13722	.15389	.17056	.18722
15	.10417	.12083	.13750	.15417	.17083	.18750
16	.10444	.12111	.13778	.15444	.17111	.18778
17	.10472	.12139	.13806	.15472	.17139	.18806
18	.10500	.12167	.13833	.15500	.17167	.18833
19	.10528	.12194	.13861	.15528	.17194	.18861
20	.10556	.12222	.13889	.15556	.17222	.18889
21	.10583	.12250	.13917	.15583	.17250	.18917
22	.10611	.12278	.13944	.15611	.17278	.18944
23	.10639	.12306	.13972	.15639	.17306	.18972
24	.10667	.12333	.14000	.15667	.17333	.19000
25	.10694	.12361	.14028	.15694	.17361	.19028
26	.10722	.12389	.14056	.15722	.17389	.19056
27	.10750	.12417	.14083	.15750	.17417	.19083
28	.10778	.12444	.14111	.15778	.17444	.19111
29	.10806	.12472	.14139	.15806	.17472	.19139
30	.10833	.12500	.14167	.15833	.17500	.19167
31	.10861	.12528	.14194	.15861	.17528	.19194
32	.10889	.12556	.14222	.15889	.17556	.19222
33	.10917	.12583	.14250	.15917	.17583	.19250
34	.10944	.12611	.14278	.15944	.17611	.19278
35	.10972	.12639	.14306	.15972	.17639	.19306
36	.11000	.12667	.14333	.16000	.17667	.19333
37	.11028	.12694	.14361	.16028	.17694	.19361
38	.11056	.12722	.14389	.16056	.17722	.19389
39	.11083	.12750	.14417	.16083	.17750	.19417
40	.11111	.12778	.14444	.16111	.17778	.19444
41	.11139	.12806	.14472	.16139	.17806	.19472
42	.11167	.12833	.14500	.16167	.17833	.19500
43	.11194	.12861	.14528	.16194	.17861	.19528
44	.11222	.12889	.14556	.16222	.17889	.19556
45	.11250	.12917	.14583	.16250	.17917	.19583
46	.11278	.12944	.14611	.16278	.17944	.19611
47	.11306	.12972	.14639	.16306	.17972	.19639
48	.11333	.13000	.14667	.16333	.18000	.19667
49	.11361	.13028	.14694	.16361	.18028	.19694
50	.11389	.13056	.14722	.16389	.18056	.19722
51	.11417	.13083	.14750	.16417	.18083	.19750
52	.11444	.13111	.14778	.16444	.18111	.19778
53	.11472	.13139	.14806	.16472	.18139	.19806
54	.11500	.13167	.14833	.16500	.18167	.19833
55	.11528	.13194	.14861	.16528	.18194	.19861
56	.11556	.13222	.14889	.16556	.18222	.19889
57	.11583	.13250	.14917	.16583	.18250	.19917
58	.11611	.13278	.14944	.16611	.18278	.19944
59	.11639	.13306	.14972	.16639	.18306	.19972
60	.11667	.13333	.15000	.16667	.18333	.20000

DECIMAL DEGREE EQUIVALENTS
OF MINUTES AND SECONDS

SEC	MINUTES					
	12	13	14	15	16	17
0	.20000	.21667	.23333	.25000	.26667	.28333
1	.20028	.21694	.23361	.25028	.26694	.28361
2	.20056	.21722	.23389	.25056	.26722	.28389
3	.20083	.21750	.23417	.25083	.26750	.28417
4	.20111	.21778	.23444	.25111	.26778	.28444
5	.20139	.21806	.23472	.25139	.26806	.28472
6	.20167	.21833	.23500	.25167	.26833	.28500
7	.20194	.21861	.23528	.25194	.26861	.28528
8	.20222	.21889	.23556	.25222	.26889	.28556
9	.20250	.21917	.23583	.25250	.26917	.28583
10	.20278	.21944	.23611	.25278	.26944	.28611
11	.20306	.21972	.23639	.25306	.26972	.28639
12	.20333	.22000	.23667	.25333	.27000	.28667
13	.20361	.22028	.23694	.25361	.27028	.28694
14	.20389	.22056	.23722	.25389	.27056	.28722
15	.20417	.22083	.23750	.25417	.27083	.28750
16	.20444	.22111	.23778	.25444	.27111	.28778
17	.20472	.22139	.23806	.25472	.27139	.28806
18	.20500	.22167	.23833	.25500	.27167	.28833
19	.20528	.22194	.23861	.25528	.27194	.28861
20	.20556	.22222	.23889	.25556	.27222	.28889
21	.20583	.22250	.23917	.25583	.27250	.28917
22	.20611	.22278	.23944	.25611	.27278	.28944
23	.20639	.22306	.23972	.25639	.27306	.28972
24	.20667	.22333	.24000	.25667	.27333	.29000
25	.20694	.22361	.24028	.25694	.27361	.29028
26	.20722	.22389	.24056	.25722	.27389	.29056
27	.20750	.22417	.24083	.25750	.27417	.29083
28	.20778	.22444	.24111	.25778	.27444	.29111
29	.20806	.22472	.24139	.25806	.27472	.29139
30	.20833	.22500	.24167	.25833	.27500	.29167
31	.20861	.22528	.24194	.25861	.27528	.29194
32	.20889	.22556	.24222	.25889	.27556	.29222
33	.20917	.22583	.24250	.25917	.27583	.29250
34	.20944	.22611	.24278	.25944	.27611	.29278
35	.20972	.22639	.24306	.25972	.27639	.29306
36	.21000	.22667	.24333	.26000	.27667	.29333
37	.21028	.22694	.24361	.26028	.27694	.29361
38	.21056	.22722	.24389	.26056	.27722	.29389
39	.21083	.22750	.24417	.26083	.27750	.29417
40	.21111	.22778	.24444	.26111	.27778	.29444
41	.21139	.22806	.24472	.26139	.27806	.29472
42	.21167	.22833	.24500	.26167	.27833	.29500
43	.21194	.22861	.24528	.26194	.27861	.29528
44	.21222	.22889	.24556	.26222	.27889	.29556
45	.21250	.22917	.24583	.26250	.27917	.29583
46	.21278	.22944	.24611	.26278	.27944	.29611
47	.21306	.22972	.24639	.26306	.27972	.29639
48	.21333	.23000	.24667	.26333	.28000	.29667
49	.21361	.23028	.24694	.26361	.28028	.29694
50	.21389	.23056	.24722	.26389	.28056	.29722
51	.21417	.23083	.24750	.26417	.28083	.29750
52	.21444	.23111	.24778	.26444	.28111	.29778
53	.21472	.23139	.24806	.26472	.28139	.29806
54	.21500	.23167	.24833	.26500	.28167	.29833
55	.21528	.23194	.24861	.26528	.28194	.29861
56	.21556	.23222	.24889	.26556	.28222	.29889
57	.21583	.23250	.24917	.26583	.28250	.29917
58	.21611	.23278	.24944	.26611	.28278	.29944
59	.21639	.23306	.24972	.26639	.28306	.29972
60	.21667	.23333	.25000	.26667	.28333	.30000

DECIMAL DEGREE EQUIVALENTS
OF MINUTES AND SECONDS

SEC	MINUTES					
	18	19	20	21	22	23
0	.30000	.31667	.33333	.35000	.36667	.38333
1	.30028	.31694	.33361	.35028	.36694	.38361
2	.30056	.31722	.33389	.35056	.36722	.38389
3	.30083	.31750	.33417	.35083	.36750	.38417
4	.30111	.31778	.33444	.35111	.36778	.38444
5	.30139	.31806	.33472	.35139	.36806	.38472
6	.30167	.31833	.33500	.35167	.36833	.38500
7	.30194	.31861	.33528	.35194	.36861	.38528
8	.30222	.31889	.33556	.35222	.36889	.38556
9	.30250	.31917	.33583	.35250	.36917	.38583
10	.30278	.31944	.33611	.35278	.36944	.38611
11	.30306	.31972	.33639	.35306	.36972	.38639
12	.30333	.32000	.33667	.35333	.37000	.38667
13	.30361	.32028	.33694	.35361	.37028	.38694
14	.30389	.32056	.33722	.35389	.37056	.38722
15	.30417	.32083	.33750	.35417	.37083	.38750
16	.30444	.32111	.33778	.35444	.37111	.38778
17	.30472	.32139	.33806	.35472	.37139	.38806
18	.30500	.32167	.33833	.35500	.37167	.38833
19	.30528	.32194	.33861	.35528	.37194	.38861
20	.30556	.32222	.33889	.35556	.37222	.38889
21	.30583	.32250	.33917	.35583	.37250	.38917
22	.30611	.32278	.33944	.35611	.37278	.38944
23	.30639	.32306	.33972	.35639	.37306	.38972
24	.30667	.32333	.34000	.35667	.37333	.39000
25	.30694	.32361	.34028	.35694	.37361	.39028
26	.30722	.32389	.34056	.35722	.37389	.39056
27	.30750	.32417	.34083	.35750	.37417	.39083
28	.30778	.32444	.34111	.35778	.37444	.39111
29	.30806	.32472	.34139	.35806	.37472	.39139
30	.30833	.32500	.34167	.35833	.37500	.39167
31	.30861	.32528	.34194	.35861	.37528	.39194
32	.30889	.32556	.34222	.35889	.37556	.39222
33	.30917	.32583	.34250	.35917	.37583	.39250
34	.30944	.32611	.34278	.35944	.37611	.39278
35	.30972	.32639	.34306	.35972	.37639	.39306
36	.31000	.32667	.34333	.36000	.37667	.39333
37	.31028	.32694	.34361	.36028	.37694	.39361
38	.31056	.32722	.34389	.36056	.37722	.39389
39	.31083	.32750	.34417	.36083	.37750	.39417
40	.31111	.32778	.34444	.36111	.37778	.39444
41	.31139	.32806	.34472	.36139	.37806	.39472
42	.31167	.32833	.34500	.36167	.37833	.39500
43	.31194	.32861	.34528	.36194	.37861	.39528
44	.31222	.32889	.34556	.36222	.37889	.39556
45	.31250	.32917	.34583	.36250	.37917	.39583
46	.31278	.32944	.34611	.36278	.37944	.39611
47	.31306	.32972	.34639	.36306	.37972	.39639
48	.31333	.33000	.34667	.36333	.38000	.39667
49	.31361	.33028	.34694	.36361	.38028	.39694
50	.31389	.33056	.34722	.36389	.38056	.39722
51	.31417	.33083	.34750	.36417	.38083	.39750
52	.31444	.33111	.34778	.36444	.38111	.39778
53	.31472	.33139	.34806	.36472	.38139	.39806
54	.31500	.33167	.34833	.36500	.38167	.39833
55	.31528	.33194	.34861	.36528	.38194	.39861
56	.31556	.33222	.34889	.36556	.38222	.39889
57	.31583	.33250	.34917	.36583	.38250	.39917
58	.31611	.33278	.34944	.36611	.38278	.39944
59	.31639	.33306	.34972	.36639	.38306	.39972
60	.31667	.33333	.35000	.36667	.38333	.40000

DECIMAL DEGREE EQUIVALENTS
OF MINUTES AND SECONDS

SEC	MINUTES					
	24	25	26	27	28	29
0	.40000	.41667	.43333	.45000	.46667	.48333
1	.40028	.41694	.43361	.45028	.46694	.48361
2	.40056	.41722	.43389	.45056	.46722	.48389
3	.40083	.41750	.43417	.45083	.46750	.48417
4	.40111	.41778	.43444	.45111	.46778	.48444
5	.40139	.41806	.43472	.45139	.46806	.48472
6	.40167	.41833	.43500	.45167	.46833	.48500
7	.40194	.41861	.43528	.45194	.46861	.48528
8	.40222	.41889	.43556	.45222	.46889	.48556
9	.40250	.41917	.43583	.45250	.46917	.48583
10	.40278	.41944	.43611	.45278	.46944	.48611
11	.40306	.41972	.43639	.45306	.46972	.48639
12	.40333	.42000	.43667	.45333	.47000	.48667
13	.40361	.42028	.43694	.45361	.47028	.48694
14	.40389	.42056	.43722	.45389	.47056	.48722
15	.40417	.42083	.43750	.45417	.47083	.48750
16	.40444	.42111	.43778	.45444	.47111	.48778
17	.40472	.42139	.43806	.45472	.47139	.48806
18	.40500	.42167	.43833	.45500	.47167	.48833
19	.40528	.42194	.43861	.45528	.47194	.48861
20	.40556	.42222	.43889	.45556	.47222	.48889
21	.40583	.42250	.43917	.45583	.47250	.48917
22	.40611	.42278	.43944	.45611	.47278	.48944
23	.40639	.42306	.43972	.45639	.47306	.48972
24	.40667	.42333	.44000	.45667	.47333	.49000
25	.40694	.42361	.44028	.45694	.47361	.49028
26	.40722	.42389	.44056	.45722	.47389	.49056
27	.40750	.42417	.44083	.45750	.47417	.49083
28	.40778	.42444	.44111	.45778	.47444	.49111
29	.40806	.42472	.44139	.45806	.47472	.49139
30	.40833	.42500	.44167	.45833	.47500	.49167
31	.40861	.42528	.44194	.45861	.47528	.49194
32	.40889	.42556	.44222	.45889	.47556	.49222
33	.40917	.42583	.44250	.45917	.47583	.49250
34	.40944	.42611	.44278	.45944	.47611	.49278
35	.40972	.42639	.44306	.45972	.47639	.49306
36	.41000	.42667	.44333	.46000	.47667	.49333
37	.41028	.42694	.44361	.46028	.47694	.49361
38	.41056	.42722	.44389	.46056	.47722	.49389
39	.41083	.42750	.44417	.46083	.47750	.49417
40	.41111	.42778	.44444	.46111	.47778	.49444
41	.41139	.42806	.44472	.46139	.47806	.49472
42	.41167	.42833	.44500	.46167	.47833	.49500
43	.41194	.42861	.44528	.46194	.47861	.49528
44	.41222	.42889	.44556	.46222	.47889	.49556
45	.41250	.42917	.44583	.46250	.47917	.49583
46	.41278	.42944	.44611	.46278	.47944	.49611
47	.41306	.42972	.44639	.46306	.47972	.49639
48	.41333	.43000	.44667	.46333	.48000	.49667
49	.41361	.43028	.44694	.46361	.48028	.49694
50	.41389	.43056	.44722	.46389	.48056	.49722
51	.41417	.43083	.44750	.46417	.48083	.49750
52	.41444	.43111	.44778	.46444	.48111	.49778
53	.41472	.43139	.44806	.46472	.48139	.49806
54	.41500	.43167	.44833	.46500	.48167	.49833
55	.41528	.43194	.44861	.46528	.48194	.49861
56	.41556	.43222	.44889	.46556	.48222	.49889
57	.41583	.43250	.44917	.46583	.48250	.49917
58	.41611	.43278	.44944	.46611	.48278	.49944
59	.41639	.43306	.44972	.46639	.48306	.49972
60	.41667	.43333	.45000	.46667	.48333	.50000

DECIMAL DEGREE EQUIVALENTS
OF MINUTES AND SECONDS

SEC	MINUTES					
	30	31	32	33	34	35
0	.50000	.51667	.53333	.55000	.56667	.58333
1	.50028	.51694	.53361	.55028	.56694	.58361
2	.50056	.51722	.53389	.55056	.56722	.58389
3	.50083	.51750	.53417	.55083	.56750	.58417
4	.50111	.51778	.53444	.55111	.56778	.58444
5	.50139	.51806	.53472	.55139	.56806	.58472
6	.50167	.51833	.53500	.55167	.56833	.58500
7	.50194	.51861	.53528	.55194	.56861	.58528
8	.50222	.51889	.53556	.55222	.56889	.58556
9	.50250	.51917	.53583	.55250	.56917	.58583
10	.50278	.51944	.53611	.55278	.56944	.58611
11	.50306	.51972	.53639	.55306	.56972	.58639
12	.50333	.52000	.53667	.55333	.57000	.58667
13	.50361	.52028	.53694	.55361	.57028	.58694
14	.50389	.52056	.53722	.55389	.57056	.58722
15	.50417	.52083	.53750	.55417	.57083	.58750
16	.50444	.52111	.53778	.55444	.57111	.58778
17	.50472	.52139	.53806	.55472	.57139	.58806
18	.50500	.52167	.53833	.55500	.57167	.58833
19	.50528	.52194	.53861	.55528	.57194	.58861
20	.50556	.52222	.53889	.55556	.57222	.58889
21	.50583	.52250	.53917	.55583	.57250	.58917
22	.50611	.52278	.53944	.55611	.57278	.58944
23	.50639	.52306	.53972	.55639	.57306	.58972
24	.50667	.52333	.54000	.55667	.57333	.59000
25	.50694	.52361	.54028	.55694	.57361	.59028
26	.50722	.52389	.54056	.55722	.57389	.59056
27	.50750	.52417	.54083	.55750	.57417	.59083
28	.50778	.52444	.54111	.55778	.57444	.59111
29	.50806	.52472	.54139	.55806	.57472	.59139
30	.50833	.52500	.54167	.55833	.57500	.59167
31	.50861	.52528	.54194	.55861	.57528	.59194
32	.50889	.52556	.54222	.55889	.57556	.59222
33	.50917	.52583	.54250	.55917	.57583	.59250
34	.50944	.52611	.54278	.55944	.57611	.59278
35	.50972	.52639	.54306	.55972	.57639	.59306
36	.51000	.52667	.54333	.56000	.57667	.59333
37	.51028	.52694	.54361	.56028	.57694	.59361
38	.51056	.52722	.54389	.56056	.57722	.59389
39	.51083	.52750	.54417	.56083	.57750	.59417
40	.51111	.52778	.54444	.56111	.57778	.59444
41	.51139	.52806	.54472	.56139	.57806	.59472
42	.51167	.52833	.54500	.56167	.57833	.59500
43	.51194	.52861	.54528	.56194	.57861	.59528
44	.51222	.52889	.54556	.56222	.57889	.59556
45	.51250	.52917	.54583	.56250	.57917	.59583
46	.51278	.52944	.54611	.56278	.57944	.59611
47	.51306	.52972	.54639	.56306	.57972	.59639
48	.51333	.53000	.54667	.56333	.58000	.59667
49	.51361	.53028	.54694	.56361	.58028	.59694
50	.51389	.53056	.54722	.56389	.58056	.59722
51	.51417	.53083	.54750	.56417	.58083	.59750
52	.51444	.53111	.54778	.56444	.58111	.59778
53	.51472	.53139	.54806	.56472	.58139	.59806
54	.51500	.53167	.54833	.56500	.58167	.59833
55	.51528	.53194	.54861	.56528	.58194	.59861
56	.51556	.53222	.54889	.56556	.58222	.59889
57	.51583	.53250	.54917	.56583	.58250	.59917
58	.51611	.53278	.54944	.56611	.58278	.59944
59	.51639	.53306	.54972	.56639	.58306	.59972
60	.51667	.53333	.55000	.56667	.58333	.60000

SEC	MINUTES					
	36	37	38	39	40	41
0	.60000	.61667	.63333	.65000	.66667	.68333
1	.60028	.61694	.63361	.65028	.66694	.68361
2	.60056	.61722	.63389	.65056	.66722	.68389
3	.60083	.61750	.63417	.65083	.66750	.68417
4	.60111	.61778	.63444	.65111	.66778	.68444
5	.60139	.61806	.63472	.65139	.66806	.68472
6	.60167	.61833	.63500	.65167	.66833	.68500
7	.60194	.61861	.63528	.65194	.66861	.68528
8	.60222	.61889	.63556	.65222	.66889	.68556
9	.60250	.61917	.63583	.65250	.66917	.68583
10	.60278	.61944	.63611	.65278	.66944	.68611
11	.60306	.61972	.63639	.65306	.66972	.68639
12	.60333	.62000	.63667	.65333	.67000	.68667
13	.60361	.62028	.63694	.65361	.67028	.68694
14	.60389	.62056	.63722	.65389	.67056	.68722
15	.60417	.62083	.63750	.65417	.67083	.68750
16	.60444	.62111	.63778	.65444	.67111	.68778
17	.60472	.62139	.63806	.65472	.67139	.68806
18	.60500	.62167	.63833	.65500	.67167	.68833
19	.60528	.62194	.63861	.65528	.67194	.68861
20	.60556	.62222	.63889	.65556	.67222	.68889
21	.60583	.62250	.63917	.65583	.67250	.68917
22	.60611	.62278	.63944	.65611	.67278	.68944
23	.60639	.62306	.63972	.65639	.67306	.68972
24	.60667	.62333	.64000	.65667	.67333	.69000
25	.60694	.62361	.64028	.65694	.67361	.69028
26	.60722	.62389	.64056	.65722	.67389	.69056
27	.60750	.62417	.64083	.65750	.67417	.69083
28	.60778	.62444	.64111	.65778	.67444	.69111
29	.60806	.62472	.64139	.65806	.67472	.69139
30	.60833	.62500	.64167	.65833	.67500	.69167
31	.60861	.62528	.64194	.65861	.67528	.69194
32	.60889	.62556	.64222	.65889	.67556	.69222
33	.60917	.62583	.64250	.65917	.67583	.69250
34	.60944	.62611	.64278	.65944	.67611	.69278
35	.60972	.62639	.64306	.65972	.67639	.69306
36	.61000	.62667	.64333	.66000	.67667	.69333
37	.61028	.62694	.64361	.66028	.67694	.69361
38	.61056	.62722	.64389	.66056	.67722	.69389
39	.61083	.62750	.64417	.66083	.67750	.69417
40	.61111	.62778	.64444	.66111	.67778	.69444
41	.61139	.62806	.64472	.66139	.67806	.69472
42	.61167	.62833	.64500	.66167	.67833	.69500
43	.61194	.62861	.64528	.66194	.67861	.69528
44	.61222	.62889	.64556	.66222	.67889	.69556
45	.61250	.62917	.64583	.66250	.67917	.69583
46	.61278	.62944	.64611	.66278	.67944	.69611
47	.61306	.62972	.64639	.66306	.67972	.69639
48	.61333	.63000	.64667	.66333	.68000	.69667
49	.61361	.63028	.64694	.66361	.68028	.69694
50	.61389	.63056	.64722	.66389	.68056	.69722
51	.61417	.63083	.64750	.66417	.68083	.69750
52	.61444	.63111	.64778	.66444	.68111	.69778
53	.61472	.63139	.64806	.66472	.68139	.69806
54	.61500	.63167	.64833	.66500	.68167	.69833
55	.61528	.63194	.64861	.66528	.68194	.69861
56	.61556	.63222	.64889	.66556	.68222	.69889
57	.61583	.63250	.64917	.66583	.68250	.69917
58	.61611	.63278	.64944	.66611	.68278	.69944
59	.61639	.63306	.64972	.66639	.68306	.69972
60	.61667	.63333	.65000	.66667	.68333	.70000

DECIMAL DEGREE EQUIVALENTS
OF MINUTES AND SECONDS

SEC	MINUTES					
	42	43	44	45	46	47
0	.70000	.71667	.73333	.75000	.76667	.78333
1	.70028	.71694	.73361	.75028	.76694	.78361
2	.70056	.71722	.73389	.75056	.76722	.78389
3	.70083	.71750	.73417	.75083	.76750	.78417
4	.70111	.71778	.73444	.75111	.76778	.78444
5	.70139	.71806	.73472	.75139	.76806	.78472
6	.70167	.71833	.73500	.75167	.76833	.78500
7	.70194	.71861	.73528	.75194	.76861	.78528
8	.70222	.71889	.73556	.75222	.76889	.78556
9	.70250	.71917	.73583	.75250	.76917	.78583
10	.70278	.71944	.73611	.75278	.76944	.78611
11	.70306	.71972	.73639	.75306	.76972	.78639
12	.70333	.72000	.73667	.75333	.77000	.78667
13	.70361	.72028	.73694	.75361	.77028	.78694
14	.70389	.72056	.73722	.75389	.77056	.78722
15	.70417	.72083	.73750	.75417	.77083	.78750
16	.70444	.72111	.73778	.75444	.77111	.78778
17	.70472	.72139	.73806	.75472	.77139	.78806
18	.70500	.72167	.73833	.75500	.77167	.78833
19	.70528	.72194	.73861	.75528	.77194	.78861
20	.70556	.72222	.73889	.75556	.77222	.78889
21	.70583	.72250	.73917	.75583	.77250	.78917
22	.70611	.72278	.73944	.75611	.77278	.78944
23	.70639	.72306	.73972	.75639	.77306	.78972
24	.70667	.72333	.74000	.75667	.77333	.79000
25	.70694	.72361	.74028	.75694	.77361	.79028
26	.70722	.72389	.74056	.75722	.77389	.79056
27	.70750	.72417	.74083	.75750	.77417	.79083
28	.70778	.72444	.74111	.75778	.77444	.79111
29	.70806	.72472	.74139	.75806	.77472	.79139
30	.70833	.72500	.74167	.75833	.77500	.79167
31	.70861	.72528	.74194	.75861	.77528	.79194
32	.70889	.72556	.74222	.75889	.77556	.79222
33	.70917	.72583	.74250	.75917	.77583	.79250
34	.70944	.72611	.74278	.75944	.77611	.79278
35	.70972	.72639	.74306	.75972	.77639	.79306
36	.71000	.72667	.74333	.76000	.77667	.79333
37	.71028	.72694	.74361	.76028	.77694	.79361
38	.71056	.72722	.74389	.76056	.77722	.79389
39	.71083	.72750	.74417	.76083	.77750	.79417
40	.71111	.72778	.74444	.76111	.77778	.79444
41	.71139	.72806	.74472	.76139	.77806	.79472
42	.71167	.72833	.74500	.76167	.77833	.79500
43	.71194	.72861	.74528	.76194	.77861	.79528
44	.71222	.72889	.74556	.76222	.77889	.79556
45	.71250	.72917	.74583	.76250	.77917	.79583
46	.71278	.72944	.74611	.76278	.77944	.79611
47	.71306	.72972	.74639	.76306	.77972	.79639
48	.71333	.73000	.74667	.76333	.78000	.79667
49	.71361	.73028	.74694	.76361	.78028	.79694
50	.71389	.73056	.74722	.76389	.78056	.79722
51	.71417	.73083	.74750	.76417	.78083	.79750
52	.71444	.73111	.74778	.76444	.78111	.79778
53	.71472	.73139	.74806	.76472	.78139	.79806
54	.71500	.73167	.74833	.76500	.78167	.79833
55	.71528	.73194	.74861	.76528	.78194	.79861
56	.71556	.73222	.74889	.76556	.78222	.79889
57	.71583	.73250	.74917	.76583	.78250	.79917
58	.71611	.73278	.74944	.76611	.78278	.79944
59	.71639	.73306	.74972	.76639	.78306	.79972
60	.71667	.73333	.75000	.76667	.78333	.80000

DECIMAL DEGREE EQUIVALENTS
OF MINUTES AND SECONDS

SEC	MINUTES					
	48	49	50	51	52	53
0	.80000	.81667	.83333	.85000	.86667	.88333
1	.80028	.81694	.83361	.85028	.86694	.88361
2	.80056	.81722	.83389	.85056	.86722	.88389
3	.80083	.81750	.83417	.85083	.86750	.88417
4	.80111	.81778	.83444	.85111	.86778	.88444
5	.80139	.81806	.83472	.85139	.86806	.88472
6	.80167	.81833	.83500	.85167	.86833	.88500
7	.80194	.81861	.83528	.85194	.86861	.88528
8	.80222	.81889	.83556	.85222	.86889	.88556
9	.80250	.81917	.83583	.85250	.86917	.88583
10	.80278	.81944	.83611	.85278	.86944	.88611
11	.80306	.81972	.83639	.85306	.86972	.88639
12	.80333	.82000	.83667	.85333	.87000	.88667
13	.80361	.82028	.83694	.85361	.87028	.88694
14	.80389	.82056	.83722	.85389	.87056	.88722
15	.80417	.82083	.83750	.85417	.87083	.88750
16	.80444	.82111	.83778	.85444	.87111	.88778
17	.80472	.82139	.83806	.85472	.87139	.88806
18	.80500	.82167	.83833	.85500	.87167	.88833
19	.80528	.82194	.83861	.85528	.87194	.88861
20	.80556	.82222	.83889	.85556	.87222	.88889
21	.80583	.82250	.83917	.85583	.87250	.88917
22	.80611	.82278	.83944	.85611	.87278	.88944
23	.80639	.82306	.83972	.85639	.87306	.88972
24	.80667	.82333	.84000	.85667	.87333	.89000
25	.80694	.82361	.84028	.85694	.87361	.89028
26	.80722	.82389	.84056	.85722	.87389	.89056
27	.80750	.82417	.84083	.85750	.87417	.89083
28	.80778	.82444	.84111	.85778	.87444	.89111
29	.80806	.82472	.84139	.85806	.87472	.89139
30	.80833	.82500	.84167	.85833	.87500	.89167
31	.80861	.82528	.84194	.85861	.87528	.89194
32	.80889	.82556	.84222	.85889	.87556	.89222
33	.80917	.82583	.84250	.85917	.87583	.89250
34	.80944	.82611	.84278	.85944	.87611	.89278
35	.80972	.82639	.84306	.85972	.87639	.89306
36	.81000	.82667	.84333	.86000	.87667	.89333
37	.81028	.82694	.84361	.86028	.87694	.89361
38	.81056	.82722	.84389	.86056	.87722	.89389
39	.81083	.82750	.84417	.86083	.87750	.89417
40	.81111	.82778	.84444	.86111	.87778	.89444
41	.81139	.82806	.84472	.86139	.87806	.89472
42	.81167	.82833	.84500	.86167	.87833	.89500
43	.81194	.82861	.84528	.86194	.87861	.89528
44	.81222	.82889	.84556	.86222	.87889	.89556
45	.81250	.82917	.84583	.86250	.87917	.89583
46	.81278	.82944	.84611	.86278	.87944	.89611
47	.81306	.82972	.84639	.86306	.87972	.89639
48	.81333	.83000	.84667	.86333	.88000	.89667
49	.81361	.83028	.84694	.86361	.88028	.89694
50	.81389	.83056	.84722	.86389	.88056	.89722
51	.81417	.83083	.84750	.86417	.88083	.89750
52	.81444	.83111	.84778	.86444	.88111	.89778
53	.81472	.83139	.84806	.86472	.88139	.89806
54	.81500	.83167	.84833	.86500	.88167	.89833
55	.81528	.83194	.84861	.86528	.88194	.89861
56	.81556	.83222	.84889	.86556	.88222	.89889
57	.81583	.83250	.84917	.86583	.88250	.89917
58	.81611	.83278	.84944	.86611	.88278	.89944
59	.81639	.83306	.84972	.86639	.88306	.89972
60	.81667	.83333	.85000	.86667	.88333	.90000

DECIMAL DEGREE EQUIVALENTS
OF MINUTES AND SECONDS

SEC	MINUTES					
	54	55	56	57	58	59
0	.90000	.91667	.93333	.95000	.96667	.98333
1	.90028	.91694	.93361	.95028	.96694	.98361
2	.90056	.91722	.93389	.95056	.96722	.98389
3	.90083	.91750	.93417	.95083	.96750	.98417
4	.90111	.91778	.93444	.95111	.96778	.98444
5	.90139	.91806	.93472	.95139	.96806	.98472
6	.90167	.91833	.93500	.95167	.96833	.98500
7	.90194	.91861	.93528	.95194	.96861	.98528
8	.90222	.91889	.93556	.95222	.96889	.98556
9	.90250	.91917	.93583	.95250	.96917	.98583
10	.90278	.91944	.93611	.95278	.96944	.98611
11	.90306	.91972	.93639	.95306	.96972	.98639
12	.90333	.92000	.93667	.95333	.97000	.98667
13	.90361	.92028	.93694	.95361	.97028	.98694
14	.90389	.92056	.93722	.95389	.97056	.98722
15	.90417	.92083	.93750	.95417	.97083	.98750
16	.90444	.92111	.93778	.95444	.97111	.98778
17	.90472	.92139	.93806	.95472	.97139	.98806
18	.90500	.92167	.93833	.95500	.97167	.98833
19	.90528	.92194	.93861	.95528	.97194	.98861
20	.90556	.92222	.93889	.95556	.97222	.98889
21	.90583	.92250	.93917	.95583	.97250	.98917
22	.90611	.92278	.93944	.95611	.97278	.98944
23	.90639	.92306	.93972	.95639	.97306	.98972
24	.90667	.92333	.94000	.95667	.97333	.99000
25	.90694	.92361	.94028	.95694	.97361	.99028
26	.90722	.92389	.94056	.95722	.97389	.99056
27	.90750	.92417	.94083	.95750	.97417	.99083
28	.90778	.92444	.94111	.95778	.97444	.99111
29	.90806	.92472	.94139	.95806	.97472	.99139
30	.90833	.92500	.94167	.95833	.97500	.99167
31	.90861	.92528	.94194	.95861	.97528	.99194
32	.90889	.92556	.94222	.95889	.97556	.99222
33	.90917	.92583	.94250	.95917	.97583	.99250
34	.90944	.92611	.94278	.95944	.97611	.99278
35	.90972	.92639	.94306	.95972	.97639	.99306
36	.91000	.92667	.94333	.96000	.97667	.99333
37	.91028	.92694	.94361	.96028	.97694	.99361
38	.91056	.92722	.94389	.96056	.97722	.99389
39	.91083	.92750	.94417	.96083	.97750	.99417
40	.91111	.92778	.94444	.96111	.97778	.99444
41	.91139	.92806	.94472	.96139	.97806	.99472
42	.91167	.92833	.94500	.96167	.97833	.99500
43	.91194	.92861	.94528	.96194	.97861	.99528
44	.91222	.92889	.94556	.96222	.97889	.99556
45	.91250	.92917	.94583	.96250	.97917	.99583
46	.91278	.92944	.94611	.96278	.97944	.99611
47	.91306	.92972	.94639	.96306	.97972	.99639
48	.91333	.93000	.94667	.96333	.98000	.99667
49	.91361	.93028	.94694	.96361	.98028	.99694
50	.91389	.93056	.94722	.96389	.98056	.99722
51	.91417	.93083	.94750	.96417	.98083	.99750
52	.91444	.93111	.94778	.96444	.98111	.99778
53	.91472	.93139	.94806	.96472	.98139	.99806
54	.91500	.93167	.94833	.96500	.98167	.99833
55	.91528	.93194	.94861	.96528	.98194	.99861
56	.91556	.93222	.94889	.96556	.98222	.99889
57	.91583	.93250	.94917	.96583	.98250	.99917
58	.91611	.93278	.94944	.96611	.98278	.99944
59	.91639	.93306	.94972	.96639	.98306	.99972
60	.91667	.93333	.95000	.96667	.98333	1.00000

S T A D I A R E D U C T I O N

HORIZONTAL DISTANCE AND DIFFERENCE IN ELEVATION
CONVERSION FACTORS FOR STADIA READINGS

ZENITH ⋌	90 ,270		91 ,271		92 ,272	
VERTICAL ⋌	0		1		2	
MINUTES	HD	DE	HD	DE	HD	DE
0 60	100.00	0	99.97	1.74	99.88	3.49
2 58	100.00	.06	99.97	1.80	99.87	3.55
4 56	100.00	.12	99.97	1.86	99.87	3.60
6 54	100.00	.17	99.96	1.92	99.87	3.66
8 52	100.00	.23	99.96	1.98	99.86	3.72
10 50	100.00	.29	99.96	2.04	99.86	3.78
12 48	100.00	.35	99.96	2.09	99.85	3.84
14 46	100.00	.41	99.95	2.15	99.85	3.89
16 44	100.00	.47	99.95	2.21	99.84	3.95
18 42	100.00	.52	99.95	2.27	99.84	4.01
20 40	100.00	.58	99.95	2.33	99.83	4.07
22 38	100.00	.64	99.94	2.38	99.83	4.13
24 36	100.00	.70	99.94	2.44	99.82	4.18
26 34	99.99	.76	99.94	2.50	99.82	4.24
28 32	99.99	.81	99.93	2.56	99.81	4.30
30 30	99.99	.87	99.93	2.62	99.81	4.36
32 28	99.99	.93	99.93	2.67	99.80	4.42
34 26	99.99	.99	99.93	2.73	99.80	4.47
36 24	99.99	1.05	99.92	2.79	99.79	4.53
38 22	99.99	1.11	99.92	2.85	99.79	4.59
40 20	99.99	1.16	99.92	2.91	99.78	4.65
42 18	99.99	1.22	99.91	2.97	99.78	4.71
44 16	99.98	1.28	99.91	3.02	99.77	4.76
46 14	99.98	1.34	99.90	3.08	99.77	4.82
48 12	99.98	1.40	99.90	3.14	99.76	4.88
50 10	99.98	1.45	99.90	3.20	99.76	4.94
52 8	99.98	1.51	99.89	3.26	99.75	4.99
54 6	99.98	1.57	99.89	3.31	99.74	5.05
56 4	99.97	1.63	99.89	3.37	99.74	5.11
58 2	99.97	1.69	99.88	3.43	99.73	5.17
60 0	99.97	1.74	99.88	3.49	99.73	5.23
ZENITH ⋌	89 ,269		88 ,268		87 ,267	
C= .75	.75	.01	.75	.02	.75	.03
C=1.00	1.00	.01	1.00	.03	1.00	.04
C=1.25	1.25	.01	1.25	.03	1.25	.05

S T A D I A   R E D U C T I O N

HORIZONTAL DISTANCE AND DIFFERENCE IN ELEVATION
CONVERSION FACTORS FOR STADIA READINGS

ZENITH ∡		93 ,273		94 ,274		95 ,275	
VERTICAL ∡		3		4		5	
MINUTES		HD	DE	HD	DE	HD	DE
0	60	99.73	5.23	99.51	6.96	99.24	8.68
2	58	99.72	5.28	99.51	7.02	99.23	8.74
4	56	99.71	5.34	99.50	7.07	99.22	8.80
6	54	99.71	5.40	99.49	7.13	99.21	8.85
8	52	99.70	5.46	99.48	7.19	99.20	8.91
10	50	99.69	5.52	99.47	7.25	99.19	8.97
12	48	99.69	5.57	99.46	7.30	99.18	9.03
14	46	99.68	5.63	99.46	7.36	99.17	9.08
16	44	99.68	5.69	99.45	7.42	99.16	9.14
18	42	99.67	5.75	99.44	7.48	99.15	9.20
20	40	99.66	5.80	99.43	7.53	99.14	9.25
22	38	99.66	5.86	99.42	7.59	99.13	9.31
24	36	99.65	5.92	99.41	7.65	99.11	9.37
26	34	99.64	5.98	99.40	7.71	99.10	9.43
28	32	99.63	6.04	99.39	7.76	99.09	9.48
30	30	99.63	6.09	99.38	7.82	99.08	9.54
32	28	99.62	6.15	99.38	7.88	99.07	9.60
34	26	99.61	6.21	99.37	7.94	99.06	9.65
36	24	99.61	6.27	99.36	7.99	99.05	9.71
38	22	99.60	6.32	99.35	8.05	99.04	9.77
40	20	99.59	6.38	99.34	8.11	99.03	9.83
42	18	99.58	6.44	99.33	8.17	99.01	9.88
44	16	99.58	6.50	99.32	8.22	99.00	9.94
46	14	99.57	6.56	99.31	8.28	98.99	10.00
48	12	99.56	6.61	99.30	8.34	98.98	10.05
50	10	99.55	6.67	99.29	8.40	98.97	10.11
52	8	99.55	6.73	99.28	8.45	98.96	10.17
54	6	99.54	6.79	99.27	8.51	98.94	10.22
56	4	99.53	6.84	99.26	8.57	98.93	10.28
58	2	99.52	6.90	99.25	8.63	98.92	10.34
60	0	99.51	6.96	99.24	8.68	98.91	10.40

ZENITH ∡		86 ,266		85 ,265		84 ,264	
C= .75		.75	.05	.75	.06	.75	.07
C=1.00		1.00	.06	1.00	.08	1.00	.10
C=1.25		1.25	.08	1.25	.10	1.24	.12

S T A D I A   R E D U C T I O N

HORIZONTAL DISTANCE AND DIFFERENCE IN ELEVATION
CONVERSION FACTORS FOR STADIA READINGS

ZENITH ∡	96 ,276		97 ,277		98 ,278	
VERTICAL ∡	6		7		8	
MINUTES	HD	DE	HD	DE	HD	DE
0 60	98.91	10.40	98.51	12.10	98.06	13.78
2 58	98.90	10.45	98.50	12.15	98.05	13.84
4 56	98.88	10.51	98.49	12.21	98.03	13.89
6 54	98.87	10.57	98.47	12.27	98.01	13.95
8 52	98.86	10.62	98.46	12.32	98.00	14.01
10 50	98.85	10.68	98.44	12.38	97.98	14.06
12 48	98.83	10.74	98.43	12.43	97.97	14.12
14 46	98.82	10.79	98.41	12.49	97.95	14.17
16 44	98.81	10.85	98.40	12.55	97.93	14.23
18 42	98.80	10.91	98.39	12.60	97.92	14.28
20 40	98.78	10.96	98.37	12.66	97.90	14.34
22 38	98.77	11.02	98.36	12.72	97.88	14.40
24 36	98.76	11.08	98.34	12.77	97.87	14.45
26 34	98.74	11.13	98.33	12.83	97.85	14.51
28 32	98.73	11.19	98.31	12.88	97.83	14.56
30 30	98.72	11.25	98.30	12.94	97.82	14.62
32 28	98.71	11.30	98.28	13.00	97.80	14.67
34 26	98.69	11.36	98.27	13.05	97.78	14.73
36 24	98.68	11.42	98.25	13.11	97.76	14.79
38 22	98.67	11.47	98.24	13.17	97.75	14.84
40 20	98.65	11.53	98.22	13.22	97.73	14.90
42 18	98.64	11.59	98.20	13.28	97.71	14.95
44 16	98.63	11.64	98.19	13.33	97.69	15.01
46 14	98.61	11.70	98.17	13.39	97.68	15.06
48 12	98.60	11.76	98.16	13.45	97.66	15.12
50 10	98.58	11.81	98.14	13.50	97.64	15.17
52 8	98.57	11.87	98.13	13.56	97.62	15.23
54 6	98.56	11.93	98.11	13.61	97.61	15.28
56 4	98.54	11.98	98.10	13.67	97.59	15.34
58 2	98.53	12.04	98.08	13.73	97.57	15.40
60 0	98.51	12.10	98.06	13.78	97.55	15.45
ZENITH ∡	83 ,263		82 ,262		81 ,261	
C= .75	.75	.08	.74	.10	.74	.11
C=1.00	.99	.11	.99	.13	.99	.15
C=1.25	1.24	.14	1.24	.16	1.24	.18

STADIA   REDUCTION

HORIZONTAL DISTANCE AND DIFFERENCE IN ELEVATION
CONVERSION FACTORS FOR STADIA READINGS

ZENITH ∡		99 ,279		100 ,280		101 ,281	
VERTICAL ∡		9		10		11	

MINUTES		HD	DE	HD	DE	HD	DE
0	60	97.55	15.45	96.98	17.10	96.36	18.73
2	58	97.53	15.51	96.96	17.16	96.34	18.78
4	56	97.52	15.56	96.94	17.21	96.32	18.84
6	54	97.50	15.62	96.92	17.26	96.29	18.89
8	52	97.48	15.67	96.90	17.32	96.27	18.95
10	50	97.46	15.73	96.88	17.37	96.25	19.00
12	48	97.44	15.78	96.86	17.43	96.23	19.05
14	46	97.43	15.84	96.84	17.48	96.21	19.11
16	44	97.41	15.89	96.82	17.54	96.18	19.16
18	42	97.39	15.95	96.80	17.59	96.16	19.21
20	40	97.37	16.00	96.78	17.65	96.14	19.27
22	38	97.35	16.06	96.76	17.70	96.12	19.32
24	36	97.33	16.11	96.74	17.76	96.09	19.38
26	34	97.31	16.17	96.72	17.81	96.07	19.43
28	32	97.29	16.22	96.70	17.86	96.05	19.48
30	30	97.28	16.28	96.68	17.92	96.03	19.54
32	28	97.26	16.33	96.66	17.97	96.00	19.59
34	26	97.24	16.39	96.64	18.03	95.98	19.64
36	24	97.22	16.44	96.62	18.08	95.96	19.70
38	22	97.20	16.50	96.60	18.14	95.93	19.75
40	20	97.18	16.55	96.57	18.19	95.91	19.80
42	18	97.16	16.61	96.55	18.24	95.89	19.86
44	16	97.14	16.66	96.53	18.30	95.86	19.91
46	14	97.12	16.72	96.51	18.35	95.84	19.96
48	12	97.10	16.77	96.49	18.41	95.82	20.02
50	10	97.08	16.83	96.47	18.46	95.79	20.07
52	8	97.06	16.88	96.45	18.51	95.77	20.12
54	6	97.04	16.94	96.42	18.57	95.75	20.18
56	4	97.02	16.99	96.40	18.62	95.72	20.23
58	2	97.00	17.05	96.38	18.68	95.70	20.28
60	0	96.98	17.10	96.36	18.73	95.68	20.34

ZENITH ∡		80 ,260		79 ,259		78 ,258	

C= .75		.74	.12	.74	.14	.73	.15
C=1.00		.99	.17	.98	.18	.98	.20
C=1.25		1.23	.21	1.23	.23	1.22	.25

S T A D I A   R E D U C T I O N

HORIZONTAL DISTANCE AND DIFFERENCE IN ELEVATION
CONVERSION FACTORS FOR STADIA READINGS

| ZENITH ∡ | | 102 ,282 | | 103 ,283 | | 104 ,284 | |
| VERTICAL ∡ | | 12 | | 13 | | 14 | |
MINUTES		HD	DE	HD	DE	HD	DE
0	60	95.68	20.34	94.94	21.92	94.15	23.47
2	58	95.65	20.39	94.91	21.97	94.12	23.52
4	56	95.63	20.44	94.89	22.02	94.09	23.58
6	54	95.61	20.50	94.86	22.08	94.07	23.63
8	52	95.58	20.55	94.84	22.13	94.04	23.68
10	50	95.56	20.60	94.81	22.18	94.01	23.73
12	48	95.53	20.66	94.79	22.23	93.98	23.78
14	46	95.51	20.71	94.76	22.28	93.95	23.83
16	44	95.49	20.76	94.73	22.34	93.93	23.88
18	42	95.46	20.81	94.71	22.39	93.90	23.93
20	40	95.44	20.87	94.68	22.44	93.87	23.99
22	38	95.41	20.92	94.65	22.49	93.84	24.04
24	36	95.39	20.97	94.63	22.54	93.82	24.09
26	34	95.36	21.03	94.60	22.60	93.79	24.14
28	32	95.34	21.08	94.58	22.65	93.76	24.19
30	30	95.32	21.13	94.55	22.70	93.73	24.24
32	28	95.29	21.18	94.52	22.75	93.70	24.29
34	26	95.27	21.24	94.50	22.80	93.67	24.34
36	24	95.24	21.29	94.47	22.85	93.65	24.39
38	22	95.22	21.34	94.44	22.91	93.62	24.44
40	20	95.19	21.39	94.42	22.96	93.59	24.49
42	18	95.17	21.45	94.39	23.01	93.56	24.55
44	16	95.14	21.50	94.36	23.06	93.53	24.60
46	14	95.12	21.55	94.34	23.11	93.50	24.65
48	12	95.09	21.60	94.31	23.16	93.47	24.70
50	10	95.07	21.66	94.28	23.22	93.45	24.75
52	8	95.04	21.71	94.26	23.27	93.42	24.80
54	6	95.02	21.76	94.23	23.32	93.39	24.85
56	4	94.99	21.81	94.20	23.37	93.36	24.90
58	2	94.97	21.87	94.17	23.42	93.33	24.95
60	0	94.94	21.92	94.15	23.47	93.30	25.00

ZENITH ∡		77 ,257		76 ,256		75 ,255	
C= .75		.73	.16	.73	.18	.73	.19
C=1.00		.98	.22	.97	.23	.97	.25
C=1.25		1.22	.27	1.22	.29	1.21	.31

S T A D I A R E D U C T I O N

HORIZONTAL DISTANCE AND DIFFERENCE IN ELEVATION
CONVERSION FACTORS FOR STADIA READINGS

ZENITH ⋠		105 ,285		106 ,286		107 ,287	
VERTICAL ⋠		15		16		17	
MINUTES		HD	DE	HD	DE	HD	DE
0	60	93.30	25.00	92.40	26.50	91.45	27.96
2	58	93.27	25.05	92.37	26.55	91.42	28.01
4	56	93.24	25.10	92.34	26.59	91.39	28.06
6	54	93.21	25.15	92.31	26.64	91.35	28.10
8	52	93.18	25.20	92.28	26.69	91.32	28.15
10	50	93.16	25.25	92.25	26.74	91.29	28.20
12	48	93.13	25.30	92.22	26.79	91.26	28.25
14	46	93.10	25.35	92.19	26.84	91.22	28.30
16	44	93.07	25.40	92.15	26.89	91.19	28.34
18	42	93.04	25.45	92.12	26.94	91.16	28.39
20	40	93.01	25.50	92.09	26.99	91.12	28.44
22	38	92.98	25.55	92.06	27.04	91.09	28.49
24	36	92.95	25.60	92.03	27.09	91.06	28.54
26	34	92.92	25.65	92.00	27.13	91.02	28.58
28	32	92.89	25.70	91.97	27.18	90.99	28.63
30	30	92.86	25.75	91.93	27.23	90.96	28.68
32	28	92.83	25.80	91.90	27.28	90.92	28.73
34	26	92.80	25.85	91.87	27.33	90.89	28.77
36	24	92.77	25.90	91.84	27.38	90.86	28.82
38	22	92.74	25.95	91.81	27.43	90.82	28.87
40	20	92.71	26.00	91.77	27.48	90.79	28.92
42	18	92.68	26.05	91.74	27.52	90.76	28.96
44	16	92.65	26.10	91.71	27.57	90.72	29.01
46	14	92.62	26.15	91.68	27.62	90.69	29.06
48	12	92.59	26.20	91.65	27.67	90.66	29.11
50	10	92.56	26.25	91.61	27.72	90.62	29.15
52	8	92.53	26.30	91.58	27.77	90.59	29.20
54	6	92.49	26.35	91.55	27.81	90.55	29.25
56	4	92.46	26.40	91.52	27.86	90.52	29.30
58	2	92.43	26.45	91.48	27.91	90.49	29.34
60	0	92.40	26.50	91.45	27.96	90.45	29.39

ZENITH ⋠		74 ,254		73 ,253		72 ,252	
C= .75		.72	.20	.72	.21	.72	.23
C=1.00		.96	.27	.96	.28	.95	.30
C=1.25		1.20	.33	1.20	.36	1.19	.38

```
               S T A D I A    R E D U C T I O N
   HORIZONTAL DISTANCE AND DIFFERENCE IN ELEVATION
          CONVERSION FACTORS FOR STADIA READINGS
```

ZENITH ⋋	108 ,288		109 ,289		110 ,290	
VERTICAL ⋋	18		19		20	
MINUTES	HD	DE	HD	DE	HD	DE
0 60	90.45	29.39	89.40	30.78	88.30	32.14
2 58	90.42	29.44	89.36	30.83	88.26	32.18
4 56	90.38	29.48	89.33	30.87	88.23	32.23
6 54	90.35	29.53	89.29	30.92	88.19	32.27
8 52	90.31	29.58	89.26	30.97	88.15	32.32
10 50	90.28	29.62	89.22	31.01	88.11	32.36
12 48	90.24	29.67	89.18	31.06	88.08	32.41
14 46	90.21	29.72	89.15	31.10	88.04	32.45
16 44	90.18	29.76	89.11	31.15	88.00	32.49
18 42	90.14	29.81	89.08	31.19	87.96	32.54
20 40	90.11	29.86	89.04	31.24	87.93	32.58
22 38	90.07	29.90	89.00	31.28	87.89	32.63
24 36	90.04	29.95	88.97	31.33	87.85	32.67
26 34	90.00	30.00	88.93	31.38	87.81	32.72
28 32	89.97	30.04	88.89	31.42	87.77	32.76
30 30	89.93	30.09	88.86	31.47	87.74	32.80
32 28	89.90	30.14	88.82	31.51	87.70	32.85
34 26	89.86	30.18	88.78	31.56	87.66	32.89
36 24	89.83	30.23	88.75	31.60	87.62	32.93
38 22	89.79	30.28	88.71	31.65	87.58	32.98
40 20	89.76	30.32	88.67	31.69	87.54	33.02
42 18	89.72	30.37	88.64	31.74	87.51	33.07
44 16	89.69	30.41	88.60	31.78	87.47	33.11
46 14	89.65	30.46	88.56	31.83	87.43	33.15
48 12	89.61	30.51	88.53	31.87	87.39	33.20
50 10	89.58	30.55	88.49	31.92	87.35	33.24
52 8	89.54	30.60	88.45	31.96	87.31	33.28
54 6	89.51	30.65	88.41	32.01	87.27	33.33
56 4	89.47	30.69	88.38	32.05	87.23	33.37
58 2	89.44	30.74	88.34	32.09	87.20	33.41
60 0	89.40	30.78	88.30	32.14	87.16	33.46
ZENITH ⋋	71 ,251		70 ,250		69 ,249	
C= .75	.71	.24	.71	.25	.70	.26
C=1.00	.95	.32	.94	.33	.94	.35
C=1.25	1.19	.40	1.18	.42	1.17	.44

S T A D I A R E D U C T I O N

HORIZONTAL DISTANCE AND DIFFERENCE IN ELEVATION
CONVERSION FACTORS FOR STADIA READINGS

| ZENITH ⋊ | 111 ,291 | | 112 ,292 | | 113 ,293 | |
| VERTICAL ⋊ | 21 | | 22 | | 23 | |
MINUTES	HD	DE	HD	DE	HD	DE
0 60	87.16	33.46	85.97	34.73	84.73	35.97
2 58	87.12	33.50	85.93	34.77	84.69	36.01
4 56	87.08	33.54	85.89	34.82	84.65	36.05
6 54	87.04	33.59	85.85	34.86	84.61	36.09
8 52	87.00	33.63	85.80	34.90	84.57	36.13
10 50	86.96	33.67	85.76	34.94	84.52	36.17
12 48	86.92	33.72	85.72	34.98	84.48	36.21
14 46	86.88	33.76	85.68	35.02	84.44	36.25
16 44	86.84	33.80	85.64	35.07	84.40	36.29
18 42	86.80	33.84	85.60	35.11	84.35	36.33
20 40	86.77	33.89	85.56	35.15	84.31	36.37
22 38	86.73	33.93	85.52	35.19	84.27	36.41
24 36	86.69	33.97	85.48	35.23	84.23	36.45
26 34	86.65	34.01	85.44	35.27	84.18	36.49
28 32	86.61	34.06	85.40	35.31	84.14	36.53
30 30	86.57	34.10	85.36	35.36	84.10	36.57
32 28	86.53	34.14	85.31	35.40	84.06	36.61
34 26	86.49	34.18	85.27	35.44	84.01	36.65
36 24	86.45	34.23	85.23	35.48	83.97	36.69
38 22	86.41	34.27	85.19	35.52	83.93	36.73
40 20	86.37	34.31	85.15	35.56	83.89	36.77
42 18	86.33	34.35	85.11	35.60	83.84	36.80
44 16	86.29	34.40	85.07	35.64	83.80	36.84
46 14	86.25	34.44	85.02	35.68	83.76	36.88
48 12	86.21	34.48	84.98	35.72	83.72	36.92
50 10	86.17	34.52	84.94	35.76	83.67	36.96
52 8	86.13	34.57	84.90	35.80	83.63	37.00
54 6	86.09	34.61	84.85	35.85	83.59	37.04
56 4	86.05	34.65	84.82	35.89	83.54	37.08
58 2	86.01	34.69	84.77	35.93	83.50	37.12
60 0	85.97	34.73	84.73	35.97	83.46	37.16

ZENITH ⋊	68 ,248		67 ,247		66 ,246	

C= .75	.70	.27	.69	.29	.69	.30
C=1.00	.93	.37	.92	.38	.92	.40
C=1.25	1.16	.46	1.15	.48	1.15	.50

```
              S T A D I A    R E D U C T I O N

    HORIZONTAL DISTANCE AND DIFFERENCE IN ELEVATION
         CONVERSION FACTORS FOR STADIA READINGS
```

ZENITH ⋜	114 ,294		115 ,295		116 ,296	
VERTICAL ⋜	24		25		26	
MINUTES	HD	DE	HD	DE	HD	DE
0 60	83.46	37.16	82.14	38.30	80.78	39.40
2 58	83.41	37.20	82.09	38.34	80.74	39.44
4 56	83.37	37.23	82.05	38.38	80.69	39.47
6 54	83.33	37.27	82.01	38.41	80.65	39.51
8 52	83.28	37.31	81.96	38.45	80.60	39.54
10 50	83.24	37.35	81.92	38.49	80.55	39.58
12 48	83.20	37.39	81.87	38.53	80.51	39.61
14 46	83.15	37.43	81.83	38.56	80.46	39.65
16 44	83.11	37.47	81.78	38.60	80.41	39.69
18 42	83.07	37.51	81.74	38.64	80.37	39.72
20 40	83.02	37.54	81.69	38.67	80.32	39.76
22 38	82.98	37.58	81.65	38.71	80.28	39.79
24 36	82.93	37.62	81.60	38.75	80.23	39.83
26 34	82.89	37.66	81.56	38.78	80.18	39.86
28 32	82.85	37.70	81.51	38.82	80.14	39.90
30 30	82.80	37.74	81.47	38.86	80.09	39.93
32 28	82.76	37.77	81.42	38.89	80.04	39.97
34 26	82.72	37.81	81.38	38.93	80.00	40.00
36 24	82.67	37.85	81.33	38.97	79.95	40.04
38 22	82.63	37.89	81.28	39.00	79.90	40.07
40 20	82.58	37.93	81.24	39.04	79.86	40.11
42 18	82.54	37.96	81.19	39.08	79.81	40.14
44 16	82.49	38.00	81.15	39.11	79.76	40.18
46 14	82.45	38.04	81.10	39.15	79.72	40.21
48 12	82.41	38.08	81.06	39.18	79.67	40.24
50 10	82.36	38.11	81.01	39.22	79.62	40.28
52 8	82.32	38.15	80.97	39.26	79.58	40.31
54 6	82.27	38.19	80.92	39.29	79.53	40.35
56 4	82.23	38.23	80.87	39.33	79.48	40.38
58 2	82.18	38.26	80.83	39.36	79.44	40.42
60 0	82.14	38.30	80.78	39.40	79.39	40.45
ZENITH ⋜	65 ,245		64 ,244		63 ,243	
C= .75	.68	.31	.68	.32	.67	.33
C=1.00	.91	.41	.90	.43	.89	.45
C=1.25	1.14	.52	1.13	.54	1.12	.56

S T A D I A   R E D U C T I O N

HORIZONTAL DISTANCE AND DIFFERENCE IN ELEVATION
CONVERSION FACTORS FOR STADIA READINGS

ZENITH ∡	117 ,297		118 ,298		119 ,299	
VERTICAL ∡	27		28		29	
MINUTES	HD	DE	HD	DE	HD	DE
0 60	79.39	40.45	77.96	41.45	76.50	42.40
2 58	79.34	40.49	77.91	41.48	76.45	42.43
4 56	79.30	40.52	77.86	41.52	76.40	42.46
6 54	79.25	40.55	77.81	41.55	76.35	42.49
8 52	79.20	40.59	77.77	41.58	76.30	42.53
10 50	79.15	40.62	77.72	41.61	76.25	42.56
12 48	79.11	40.66	77.67	41.65	76.20	42.59
14 46	79.06	40.69	77.62	41.68	76.15	42.62
16 44	79.01	40.72	77.57	41.71	76.10	42.65
18 42	78.96	40.76	77.52	41.74	76.05	42.68
20 40	78.92	40.79	77.48	41.77	76.00	42.71
22 38	78.87	40.82	77.43	41.81	75.95	42.74
24 36	78.82	40.86	77.38	41.84	75.90	42.77
26 34	78.77	40.89	77.33	41.87	75.85	42.80
28 32	78.73	40.92	77.28	41.90	75.80	42.83
30 30	78.68	40.96	77.23	41.93	75.75	42.86
32 28	78.63	40.99	77.18	41.97	75.70	42.89
34 26	78.58	41.02	77.13	42.00	75.65	42.92
36 24	78.54	41.06	77.09	42.03	75.60	42.95
38 22	78.49	41.09	77.04	42.06	75.55	42.98
40 20	78.44	41.12	76.99	42.09	75.50	43.01
42 18	78.39	41.16	76.94	42.12	75.45	43.04
44 16	78.34	41.19	76.89	42.15	75.40	43.07
46 14	78.30	41.22	76.84	42.19	75.35	43.10
48 12	78.25	41.26	76.79	42.22	75.30	43.13
50 10	78.20	41.29	76.74	42.25	75.25	43.16
52 8	78.15	41.32	76.69	42.28	75.20	43.18
54 6	78.10	41.35	76.64	42.31	75.15	43.21
56 4	78.06	41.39	76.59	42.34	75.10	43.24
58 2	78.01	41.42	76.55	42.37	75.05	43.27
60 0	77.96	41.45	76.50	42.40	75.00	43.30

ZENITH ∡	62 ,242		61 ,241		60 ,240	
C= .75	.67	.35	.66	.36	.65	.37
C=1.00	.89	.46	.88	.48	.87	.49
C=1.25	1.11	.58	1.10	.60	1.09	.62

SLOPE REDUCTION TABLE FOR THE PERCENT ABNEY

SD	1% (00° 34')		2% (01° 09')		3% (01° 43')	
	HD	DE	HD	DE	HD	DE
35	35.00	.35	34.99	.70	34.98	1.05
40	40.00	.40	39.99	.80	39.98	1.20
45	45.00	.45	44.99	.90	44.98	1.35
50	50.00	.50	49.99	1.00	49.98	1.50
55	55.00	.55	54.99	1.10	54.98	1.65
60	60.00	.60	59.99	1.20	59.97	1.80
65	65.00	.65	64.99	1.30	64.97	1.95
70	70.00	.70	69.99	1.40	69.97	2.10
75	75.00	.75	74.99	1.50	74.97	2.25
80	80.00	.80	79.98	1.60	79.96	2.40
85	85.00	.85	84.98	1.70	84.96	2.55
90	90.00	.90	89.98	1.80	89.96	2.70
95	95.00	.95	94.98	1.90	94.96	2.85
100	100.00	1.00	99.98	2.00	99.96	3.00
105	104.99	1.05	104.98	2.10	104.95	3.15
110	109.99	1.10	109.98	2.20	109.95	3.30
115	114.99	1.15	114.98	2.30	114.95	3.45
120	119.99	1.20	119.98	2.40	119.95	3.60
125	124.99	1.25	124.98	2.50	124.94	3.75
130	129.99	1.30	129.97	2.60	129.94	3.90
135	134.99	1.35	134.97	2.70	134.94	4.05
140	139.99	1.40	139.97	2.80	139.94	4.20
145	144.99	1.45	144.97	2.90	144.93	4.35
150	149.99	1.50	149.97	3.00	149.93	4.50
155	154.99	1.55	154.97	3.10	154.93	4.65
160	159.99	1.60	159.97	3.20	159.93	4.80
165	164.99	1.65	164.97	3.30	164.93	4.95
170	169.99	1.70	169.97	3.40	169.92	5.10
175	174.99	1.75	174.97	3.50	174.92	5.25
180	179.99	1.80	179.96	3.60	179.92	5.40
185	184.99	1.85	184.96	3.70	184.92	5.55
190	189.99	1.90	189.96	3.80	189.91	5.70
195	194.99	1.95	194.96	3.90	194.91	5.85
200	199.99	2.00	199.96	4.00	199.91	6.00
205	204.99	2.05	204.96	4.10	204.91	6.15
210	209.99	2.10	209.96	4.20	209.91	6.30
215	214.99	2.15	214.96	4.30	214.90	6.45
220	219.99	2.20	219.96	4.40	219.90	6.60
225	224.99	2.25	224.96	4.50	224.90	6.75
230	229.99	2.30	229.95	4.60	229.90	6.90
235	234.99	2.35	234.95	4.70	234.89	7.05
240	239.99	2.40	239.95	4.80	239.89	7.20
245	244.99	2.45	244.95	4.90	244.89	7.35
250	249.99	2.50	249.95	5.00	249.89	7.50
255	254.99	2.55	254.95	5.10	254.89	7.65
260	259.99	2.60	259.95	5.20	259.88	7.80
265	264.99	2.65	264.95	5.30	264.88	7.95
270	269.99	2.70	269.95	5.40	269.88	8.10
275	274.99	2.75	274.95	5.50	274.88	8.25
280	279.99	2.80	279.94	5.60	279.87	8.40
285	284.99	2.85	284.94	5.70	284.87	8.55
290	289.99	2.90	289.94	5.80	289.87	8.70
295	294.99	2.95	294.94	5.90	294.87	8.85
300	299.99	3.00	299.94	6.00	299.87	9.00

SLOPE REDUCTION TABLE FOR THE PERCENT ABNEY

SD	4% (02° 17′) HD	DE	5% (02° 52′) HD	DE	6% (03° 26′) HD	DE
35	34.97	1.40	34.96	1.75	34.94	2.10
40	39.97	1.60	39.95	2.00	39.93	2.40
45	44.96	1.80	44.94	2.25	44.92	2.70
50	49.96	2.00	49.94	2.50	49.91	2.99
55	54.96	2.20	54.93	2.75	54.90	3.29
60	59.95	2.40	59.93	3.00	59.89	3.59
65	64.95	2.60	64.92	3.25	64.88	3.89
70	69.94	2.80	69.91	3.50	69.87	4.19
75	74.94	3.00	74.91	3.75	74.87	4.49
80	79.94	3.20	79.90	4.00	79.86	4.79
85	84.93	3.40	84.89	4.24	84.85	5.09
90	89.93	3.60	89.89	4.49	89.84	5.39
95	94.92	3.80	94.88	4.74	94.83	5.69
100	99.92	4.00	99.88	4.99	99.82	5.99
105	104.92	4.20	104.87	5.24	104.81	6.29
110	109.91	4.40	109.86	5.49	109.80	6.59
115	114.91	4.60	114.86	5.74	114.79	6.89
120	119.90	4.80	119.85	5.99	119.78	7.19
125	124.90	5.00	124.84	6.24	124.78	7.49
130	129.90	5.20	129.84	6.49	129.77	7.79
135	134.89	5.40	134.83	6.74	134.76	8.09
140	139.89	5.60	139.83	6.99	139.75	8.38
145	144.88	5.80	144.82	7.24	144.74	8.68
150	149.88	6.00	149.81	7.49	149.73	8.98
155	154.88	6.20	154.81	7.74	154.72	9.28
160	159.87	6.39	159.80	7.99	159.71	9.58
165	164.87	6.59	164.79	8.24	164.70	9.88
170	169.86	6.79	169.79	8.49	169.69	10.18
175	174.86	6.99	174.78	8.74	174.69	10.48
180	179.86	7.19	179.78	8.99	179.68	10.78
185	184.85	7.39	184.77	9.24	184.67	11.08
190	189.85	7.59	189.76	9.49	189.66	11.38
195	194.84	7.79	194.76	9.74	194.65	11.68
200	199.84	7.99	199.75	9.99	199.64	11.98
205	204.84	8.19	204.74	10.24	204.63	12.28
210	209.83	8.39	209.74	10.49	209.62	12.58
215	214.83	8.59	214.73	10.74	214.61	12.88
220	219.82	8.79	219.73	10.99	219.61	13.18
225	224.82	8.99	224.72	11.24	224.60	13.48
230	229.82	9.19	229.71	11.49	229.59	13.78
235	234.81	9.39	234.71	11.74	234.58	14.07
240	239.81	9.59	239.70	11.99	239.57	14.37
245	244.80	9.79	244.69	12.23	244.56	14.67
250	249.80	9.99	249.69	12.48	249.55	14.97
255	254.80	10.19	254.68	12.73	254.54	15.27
260	259.79	10.39	259.68	12.98	259.53	15.57
265	264.79	10.59	264.67	13.23	264.52	15.87
270	269.78	10.79	269.66	13.48	269.52	16.17
275	274.78	10.99	274.66	13.73	274.51	16.47
280	279.78	11.19	279.65	13.98	279.50	16.77
285	284.77	11.39	284.64	14.23	284.49	17.07
290	289.77	11.59	289.64	14.48	289.48	17.37
295	294.76	11.79	294.63	14.73	294.47	17.67
300	299.76	11.99	299.63	14.98	299.46	17.97

SLOPE REDUCTION TABLE FOR THE PERCENT ABNEY

SD	7%(04°00′)		8%(04°34′)		9%(05°09′)	
	HD	DE	HD	DE	HD	DE
35	34.91	2.44	34.89	2.79	34.86	3.14
40	39.90	2.79	39.87	3.19	39.84	3.59
45	44.89	3.14	44.86	3.59	44.82	4.03
50	49.88	3.49	49.84	3.99	49.80	4.48
55	54.87	3.84	54.82	4.39	54.78	4.93
60	59.85	4.19	59.81	4.78	59.76	5.38
65	64.84	4.54	64.79	5.18	64.74	5.83
70	69.83	4.89	69.78	5.58	69.72	6.27
75	74.82	5.24	74.76	5.98	74.70	6.72
80	79.80	5.59	79.75	6.38	79.68	7.17
85	84.79	5.94	84.73	6.78	84.66	7.62
90	89.78	6.28	89.71	7.18	89.64	8.07
95	94.77	6.63	94.70	7.58	94.62	8.52
100	99.76	6.98	99.68	7.97	99.60	8.96
105	104.74	7.33	104.67	8.37	104.58	9.41
110	109.73	7.68	109.65	8.77	109.56	9.86
115	114.72	8.03	114.63	9.17	114.54	10.31
120	119.71	8.38	119.62	9.57	119.52	10.76
125	124.69	8.73	124.60	9.97	124.50	11.20
130	129.68	9.08	129.59	10.37	129.48	11.65
135	134.67	9.43	134.57	10.77	134.46	12.10
140	139.66	9.78	139.55	11.16	139.44	12.55
145	144.65	10.13	144.54	11.56	144.42	13.00
150	149.63	10.47	149.52	11.96	149.40	13.45
155	154.62	10.82	154.51	12.36	154.38	13.89
160	159.61	11.17	159.49	12.76	159.36	14.34
165	164.60	11.52	164.47	13.16	164.34	14.79
170	169.59	11.87	169.46	13.56	169.32	15.24
175	174.57	12.22	174.44	13.96	174.30	15.69
180	179.56	12.57	179.43	14.35	179.28	16.13
185	184.55	12.92	184.41	14.75	184.26	16.58
190	189.54	13.27	189.39	15.15	189.24	17.03
195	194.52	13.62	194.38	15.55	194.22	17.48
200	199.51	13.97	199.36	15.95	199.19	17.93
205	204.50	14.31	204.35	16.35	204.17	18.38
210	209.49	14.66	209.33	16.75	209.15	18.82
215	214.48	15.01	214.32	17.15	214.13	19.27
220	219.46	15.36	219.30	17.54	219.11	19.72
225	224.45	15.71	224.28	17.94	224.09	20.17
230	229.44	16.06	229.27	18.34	229.07	20.62
235	234.43	16.41	234.25	18.74	234.05	21.06
240	239.41	16.76	239.24	19.14	239.03	21.51
245	244.40	17.11	244.22	19.54	244.01	21.96
250	249.39	17.46	249.20	19.94	248.99	22.41
255	254.38	17.81	254.19	20.34	253.97	22.86
260	259.37	18.16	259.17	20.73	258.95	23.31
265	264.35	18.50	264.16	21.13	263.93	23.75
270	269.34	18.85	269.14	21.53	268.91	24.20
275	274.33	19.20	274.12	21.93	273.89	24.65
280	279.32	19.55	279.11	22.33	278.87	25.10
285	284.30	19.90	284.09	22.73	283.85	25.55
290	289.29	20.25	289.08	23.13	288.83	25.99
295	294.28	20.60	294.06	23.52	293.81	26.44
300	299.27	20.95	299.04	23.92	298.79	26.89

SLOPE REDUCTION TABLE FOR THE PERCENT ABNEY

SD	10%(05° 43')		11%(06° 17')		12%(06° 51')	
	HD	DE	HD	DE	HD	DE
35	34.83	3.48	34.79	3.83	34.75	4.17
40	39.80	3.98	39.76	4.37	39.72	4.77
45	44.78	4.48	44.73	4.92	44.68	5.36
50	49.75	4.98	49.70	5.47	49.64	5.96
55	54.73	5.47	54.67	6.01	54.61	6.55
60	59.70	5.97	59.64	6.56	59.57	7.15
65	64.68	6.47	64.61	7.11	64.54	7.74
70	69.65	6.97	69.58	7.65	69.50	8.34
75	74.63	7.46	74.55	8.20	74.47	8.94
80	79.60	7.96	79.52	8.75	79.43	9.53
85	84.58	8.46	84.49	9.29	84.39	10.13
90	89.55	8.96	89.46	9.84	89.36	10.72
95	94.53	9.45	94.43	10.39	94.32	11.32
100	99.50	9.95	99.40	10.93	99.29	11.91
105	104.48	10.45	104.37	11.48	104.25	12.51
110	109.45	10.95	109.34	12.03	109.22	13.11
115	114.43	11.44	114.31	12.57	114.18	13.70
120	119.40	11.94	119.28	13.12	119.15	14.30
125	124.38	12.44	124.25	13.67	124.11	14.89
130	129.35	12.94	129.22	14.21	129.07	15.49
135	134.33	13.43	134.19	14.76	134.04	16.08
140	139.31	13.93	139.16	15.31	139.00	16.68
145	144.28	14.43	144.13	15.85	143.97	17.28
150	149.26	14.93	149.10	16.40	148.93	17.87
155	154.23	15.42	154.07	16.95	153.90	18.47
160	159.21	15.92	159.04	17.49	158.86	19.06
165	164.18	16.42	164.01	18.04	163.82	19.66
170	169.16	16.92	168.98	18.59	168.79	20.25
175	174.13	17.41	173.95	19.13	173.75	20.85
180	179.11	17.91	178.92	19.68	178.72	21.45
185	184.08	18.41	183.89	20.23	183.68	22.04
190	189.06	18.91	188.86	20.77	188.65	22.64
195	194.03	19.40	193.83	21.32	193.61	23.23
200	199.01	19.90	198.80	21.87	198.58	23.83
205	203.98	20.40	203.77	22.41	203.54	24.42
210	208.96	20.90	208.74	22.96	208.50	25.02
215	213.93	21.39	213.71	23.51	213.47	25.62
220	218.91	21.89	218.68	24.05	218.43	26.21
225	223.88	22.39	223.65	24.60	223.40	26.81
230	228.86	22.89	228.62	25.15	228.36	27.40
235	233.83	23.38	233.59	25.70	233.33	28.00
240	238.81	23.88	238.56	26.24	238.29	28.59
245	243.78	24.38	243.53	26.79	243.25	29.19
250	248.76	24.88	248.50	27.34	248.22	29.79
255	253.73	25.37	253.47	27.88	253.18	30.38
260	258.71	25.87	258.44	28.43	258.15	30.98
265	263.68	26.37	263.41	28.98	263.11	31.57
270	268.66	26.87	268.38	29.52	268.08	32.17
275	273.64	27.36	273.35	30.07	273.04	32.76
280	278.61	27.86	278.32	30.62	278.01	33.36
285	283.59	28.36	283.29	31.16	282.97	33.96
290	288.56	28.86	288.26	31.71	287.93	34.55
295	293.54	29.35	293.23	32.26	292.90	35.15
300	298.51	29.85	298.20	32.80	297.86	35.74

SLOPE REDUCTION TABLE FOR THE PERCENT ABNEY

SD	13% (07° 24')		14% (07° 58')		15% (08° 32')	
	HD	DE	HD	DE	HD	DE
35	34.71	4.51	34.66	4.85	34.61	5.19
40	39.67	5.16	39.61	5.55	39.56	5.93
45	44.62	5.80	44.57	6.24	44.50	6.68
50	49.58	6.45	49.52	6.93	49.45	7.42
55	54.54	7.09	54.47	7.63	54.39	8.16
60	59.50	7.73	59.42	8.32	59.34	8.90
65	64.46	8.38	64.37	9.01	64.28	9.64
70	69.42	9.02	69.32	9.71	69.23	10.38
75	74.37	9.67	74.28	10.40	74.17	11.13
80	79.33	10.31	79.23	11.09	79.11	11.87
85	84.29	10.96	84.18	11.79	84.06	12.61
90	89.25	11.60	89.13	12.48	89.00	13.35
95	94.21	12.25	94.08	13.17	93.95	14.09
100	99.17	12.89	99.03	13.86	98.89	14.83
105	104.12	13.54	103.99	14.56	103.84	15.58
110	109.08	14.18	108.94	15.25	108.78	16.32
115	114.04	14.93	113.89	15.94	113.73	17.06
120	119.00	15.47	118.84	16.64	118.67	17.80
125	123.96	16.11	123.79	17.33	123.62	18.54
130	128.92	16.76	128.74	18.02	128.56	19.28
135	133.87	17.40	133.70	18.72	133.51	20.03
140	138.83	18.05	138.65	19.41	138.45	20.77
145	143.79	18.69	143.60	20.10	143.40	21.51
150	148.75	19.34	148.55	20.80	148.34	22.25
155	153.71	19.98	153.50	21.49	153.29	22.99
160	158.66	20.63	158.45	22.18	158.23	23.73
165	163.62	21.27	163.41	22.88	163.17	24.48
170	168.58	21.92	168.36	23.57	168.12	25.22
175	173.54	22.56	173.31	24.26	173.06	25.96
180	178.50	23.20	178.26	24.96	178.01	26.70
185	183.46	23.85	183.21	25.65	182.95	27.44
190	188.41	24.49	188.16	26.34	187.90	28.18
195	193.37	25.14	193.12	27.04	192.84	28.93
200	198.33	25.78	198.07	27.73	197.79	29.67
205	203.29	26.43	203.02	28.42	202.73	30.41
210	208.25	27.07	207.97	29.12	207.68	31.15
215	213.21	27.72	212.92	29.81	212.62	31.89
220	218.16	28.36	217.88	30.50	217.57	32.63
225	223.12	29.01	222.83	31.20	222.51	33.38
230	228.08	29.65	227.78	31.89	227.46	34.12
235	233.04	30.30	232.73	32.58	232.40	34.86
240	238.00	30.94	237.68	33.28	237.34	35.60
245	242.96	31.58	242.63	33.97	242.29	36.34
250	247.91	32.23	247.59	34.66	247.23	37.09
255	252.87	32.87	252.54	35.36	252.18	37.83
260	257.83	33.52	257.49	36.05	257.12	38.57
265	262.79	34.16	262.44	36.74	262.07	39.31
270	267.75	34.81	267.39	37.43	267.01	40.05
275	272.71	35.45	272.34	38.13	271.96	40.79
280	277.66	36.10	277.30	38.82	276.90	41.54
285	282.62	36.74	282.25	39.51	281.85	42.28
290	287.58	37.39	287.20	40.21	286.79	43.02
295	292.54	38.03	292.15	40.90	291.74	43.76
300	297.50	38.67	297.10	41.59	296.68	44.50

SLOPE REDUCTION TABLE FOR THE PERCENT ABNEY

	16%(09°05')		17%(09°39')		18%(10°12')	
SD	HD	DE	HD	DE	HD	DE
35	34.56	5.53	34.50	5.87	34.45	6.20
40	39.50	6.32	39.43	6.70	39.37	7.09
45	44.43	7.11	44.36	7.54	44.29	7.97
50	49.37	7.90	49.29	8.38	49.21	8.86
55	54.31	8.69	54.22	9.22	54.13	9.74
60	59.25	9.48	59.15	10.06	59.05	10.63
65	64.18	10.27	64.08	10.89	63.97	11.51
70	69.12	11.06	69.01	11.73	68.89	12.40
75	74.06	11.85	73.94	12.57	73.81	13.29
80	79.00	12.64	78.87	13.41	78.73	14.17
85	83.93	13.43	83.80	14.25	83.66	15.06
90	88.87	14.22	88.73	15.08	88.58	15.94
95	93.81	15.01	93.66	15.92	93.50	16.83
100	98.74	15.80	98.59	16.76	98.42	17.72
105	103.68	16.59	103.51	17.60	103.34	18.60
110	108.62	17.38	108.44	18.44	108.26	19.49
115	113.56	18.17	113.37	19.27	113.18	20.37
120	118.49	18.96	118.30	20.11	118.10	21.26
125	123.43	19.75	123.23	20.95	123.02	22.14
130	128.37	20.54	128.16	21.79	127.94	23.03
135	133.30	21.33	133.09	22.63	132.86	23.92
140	138.24	22.12	138.02	23.46	137.79	24.80
145	143.18	22.91	142.95	24.30	142.71	25.69
150	148.12	23.70	147.88	25.14	147.63	26.57
155	153.05	24.49	152.81	25.98	152.55	27.46
160	157.99	25.28	157.74	26.82	157.47	28.34
165	162.93	26.07	162.67	27.65	162.39	29.23
170	167.86	26.86	167.60	28.49	167.31	30.12
175	172.80	27.65	172.52	29.33	172.23	31.00
180	177.74	28.44	177.45	30.17	177.15	31.89
185	182.68	29.23	182.38	31.01	182.07	32.77
190	187.61	30.02	187.31	31.84	186.99	33.66
195	192.55	30.81	192.24	32.58	191.92	34.54
200	197.49	31.60	197.17	33.52	196.84	35.43
205	202.43	32.39	202.10	34.36	201.76	36.32
210	207.36	33.18	207.03	35.20	206.68	37.20
215	212.30	33.97	211.96	36.03	211.60	38.09
220	217.24	34.76	216.89	36.87	216.52	38.97
225	222.17	35.55	221.82	37.71	221.44	39.86
230	227.11	36.34	226.75	38.55	226.36	40.75
235	232.05	37.13	231.68	39.38	231.28	41.63
240	236.99	37.92	236.61	40.22	236.20	42.52
245	241.92	38.71	241.53	41.06	241.12	43.40
250	246.86	39.50	246.46	41.90	246.05	44.29
255	251.80	40.29	251.39	42.74	250.97	45.17
260	256.73	41.08	256.32	43.57	255.89	46.06
265	261.67	41.87	261.25	44.41	260.81	46.95
270	266.61	42.66	266.18	45.25	265.73	47.83
275	271.55	43.45	271.11	46.09	270.65	48.72
280	276.48	44.24	276.04	46.93	275.57	49.60
285	281.42	45.03	280.97	47.76	280.49	50.49
290	286.36	45.82	285.90	48.60	285.41	51.37
295	291.29	46.61	290.83	49.44	290.33	52.26
300	296.23	47.40	295.76	50.28	295.25	53.15

236

SLOPE REDUCTION TABLE FOR THE PERCENT ABNEY

SD	19%(10° 45')		20%(11° 19')		21%(11° 52')	
	HD	DE	HD	DE	HD	DE
35	34.38	6.53	34.32	6.86	34.25	7.19
40	39.30	7.47	39.22	7.84	39.15	8.22
45	44.21	8.40	44.13	8.83	44.04	9.25
50	49.12	9.33	49.03	9.81	48.93	10.28
55	54.03	10.27	53.93	10.79	53.83	11.30
60	58.95	11.20	58.83	11.77	58.72	12.33
65	63.86	12.13	63.74	12.75	63.61	13.36
70	68.77	13.07	68.64	13.73	68.51	14.39
75	73.68	14.00	73.54	14.71	73.40	15.41
80	78.59	14.93	78.45	15.69	78.29	16.44
85	83.51	15.87	83.35	16.67	83.19	17.47
90	88.42	16.80	88.25	17.65	88.09	18.50
95	93.33	17.73	93.16	18.63	92.97	19.52
100	98.24	18.67	98.06	19.61	97.87	20.55
105	103.15	19.60	102.96	20.59	102.76	21.58
110	108.07	20.53	107.86	21.57	107.65	22.61
115	112.98	21.47	112.77	22.55	112.55	23.63
120	117.89	22.40	117.67	23.53	117.44	24.66
125	122.80	23.33	122.57	24.51	122.33	25.69
130	127.72	24.27	127.48	25.50	127.22	26.72
135	132.63	25.20	132.38	26.48	132.12	27.74
140	137.54	26.13	137.28	27.46	137.01	28.77
145	142.45	27.07	142.18	28.44	141.90	29.80
150	147.36	28.00	147.09	29.42	146.80	30.83
155	152.28	28.93	151.99	30.40	151.69	31.86
160	157.19	29.87	156.89	31.38	156.58	32.88
165	162.10	30.80	161.80	32.36	161.48	33.91
170	167.01	31.73	166.70	33.34	166.37	34.94
175	171.92	32.67	171.60	34.32	171.26	35.97
180	176.84	33.60	176.50	35.30	176.16	36.99
185	181.75	34.53	181.41	36.28	181.05	38.02
190	186.66	35.47	186.31	37.26	185.94	39.05
195	191.57	36.40	191.21	38.24	190.84	40.08
200	196.48	37.33	196.12	39.22	195.73	41.10
205	201.40	38.27	201.02	40.20	200.62	42.13
210	206.31	39.20	205.92	41.18	205.52	43.16
215	211.22	40.13	210.82	42.16	210.41	44.19
220	216.13	41.07	215.73	43.15	215.30	45.21
225	221.05	42.00	220.63	44.13	220.20	46.24
230	225.96	42.93	225.53	45.11	225.09	47.27
235	230.87	43.87	230.44	46.09	229.98	48.30
240	235.78	44.80	235.34	47.07	234.88	49.32
245	240.69	45.73	240.24	48.05	239.77	50.35
250	245.61	46.67	245.15	49.03	244.66	51.38
255	250.52	47.60	250.05	50.01	249.56	52.41
260	255.43	48.53	254.95	50.99	254.45	53.43
265	260.34	49.47	259.85	51.97	259.34	54.46
270	265.25	50.40	264.76	52.95	264.24	55.49
275	270.17	51.33	269.66	53.93	269.13	56.52
280	275.08	52.26	274.56	54.91	274.02	57.54
285	279.99	53.20	279.47	55.89	278.92	58.57
290	284.90	54.13	284.37	56.87	283.81	59.60
295	289.82	55.06	289.27	57.85	288.70	60.63
300	294.73	56.00	294.17	58.83	293.60	61.66

SLOPE REDUCTION TABLE FOR THE PERCENT ABNEY

SD	22%(12°24′) HD	22%(12°24′) DE	23%(12°57′) HD	23%(12°57′) DE	24%(13°30′) HD	24%(13°30′) DE
35	34.18	7.52	34.11	7.85	34.03	8.17
40	39.07	8.59	38.98	8.97	38.90	9.33
45	43.95	9.67	43.85	10.09	43.76	10.50
50	48.83	10.74	48.73	11.21	48.62	11.67
55	53.72	11.82	53.60	12.33	53.48	12.84
60	58.60	12.89	58.47	13.45	58.34	14.00
65	63.48	13.97	63.35	14.57	63.21	15.17
70	68.37	15.04	68.22	15.69	68.07	16.34
75	73.25	16.11	73.09	16.81	72.93	17.50
80	78.13	17.19	77.96	17.93	77.79	18.67
85	83.01	18.26	82.84	19.05	82.65	19.84
90	87.90	19.34	87.71	20.17	87.51	21.00
95	92.78	20.41	92.58	21.29	92.38	22.17
100	97.66	21.49	97.46	22.41	97.24	23.34
105	102.55	22.56	102.33	23.54	102.10	24.50
110	107.43	23.63	107.20	24.66	106.96	25.67
115	112.31	24.71	112.07	25.78	111.82	26.84
120	117.20	25.78	116.95	26.90	116.69	28.00
125	122.08	26.86	121.82	28.02	121.55	29.17
130	126.96	27.93	126.69	29.14	126.41	30.34
135	131.85	29.01	131.56	30.26	131.27	31.51
140	136.73	30.08	136.44	31.38	136.13	32.67
145	141.61	31.15	141.31	32.50	141.00	33.84
150	146.50	32.23	146.18	33.62	145.86	35.01
155	151.38	33.30	151.06	34.74	150.72	36.17
160	156.26	34.38	155.93	35.86	155.58	37.34
165	161.15	35.45	160.80	36.98	160.44	38.51
170	166.03	36.53	165.67	38.11	165.31	39.67
175	170.91	37.60	170.55	39.23	170.17	40.84
180	175.80	38.68	175.42	40.35	175.03	42.01
185	180.68	39.75	180.29	41.47	179.89	43.17
190	185.56	40.82	185.17	42.59	184.75	44.34
195	190.45	41.90	190.04	43.71	189.62	45.51
200	195.33	42.97	194.91	44.83	194.48	46.67
205	200.21	44.05	199.78	45.95	199.34	47.84
210	205.10	45.12	204.66	47.07	204.20	49.01
215	209.98	46.20	209.53	48.19	209.06	50.18
220	214.86	47.27	214.40	49.31	213.93	51.34
225	219.75	48.34	219.27	50.43	218.79	52.51
230	224.63	49.42	224.15	51.55	223.65	53.68
235	229.51	50.49	229.02	52.67	228.51	54.84
240	234.39	51.57	233.89	53.80	233.37	56.01
245	239.28	52.64	238.77	54.92	238.23	57.18
250	244.16	53.72	243.64	56.04	243.10	58.34
255	249.04	54.79	248.51	57.16	247.96	59.51
260	253.93	55.86	253.38	58.28	252.82	60.68
265	258.81	56.94	258.26	59.40	257.68	61.84
270	263.69	58.01	263.13	60.52	262.54	63.01
275	268.58	59.09	268.00	61.64	267.41	64.18
280	273.46	60.16	272.88	62.76	272.27	65.34
285	278.34	61.24	277.75	63.88	277.13	66.51
290	283.23	62.31	282.62	65.00	281.99	67.68
295	288.11	63.38	287.49	66.12	286.85	68.85
300	293.00	64.46	292.37	67.24	291.72	70.01

SLOPE REDUCTION TABLE FOR THE PERCENT ABNEY

SD	25% (14°02′)		26% (14°34′)		27% (15°07′)	
	HD	DE	HD	DE	HD	DE
35	33.95	8.49	33.87	8.81	33.79	9.12
40	38.81	9.70	38.71	10.07	38.62	10.43
45	43.66	10.91	43.55	11.32	43.44	11.73
50	48.51	12.13	48.39	12.58	48.27	13.03
55	53.36	13.34	53.23	13.84	53.10	14.34
60	58.21	14.55	58.07	15.10	57.93	15.64
65	63.06	15.76	62.91	16.36	62.75	16.94
70	67.91	16.98	67.75	17.61	67.58	18.25
75	72.76	18.19	72.59	18.87	72.41	19.55
80	77.61	19.40	77.43	20.13	77.23	20.85
85	82.46	20.62	82.26	21.39	82.06	22.16
90	87.31	21.83	87.10	22.65	86.89	23.46
95	92.16	23.04	91.94	23.91	91.72	24.76
100	97.01	24.25	96.78	25.16	96.54	26.07
105	101.86	25.47	101.62	26.42	101.37	27.37
110	106.72	26.68	106.46	27.68	106.20	28.67
115	111.57	27.89	111.30	28.94	111.02	29.98
120	116.42	29.10	116.14	30.20	115.85	31.28
125	121.27	30.32	120.98	31.45	120.68	32.58
130	126.12	31.53	125.82	32.71	125.51	33.89
135	130.97	32.74	130.66	33.97	130.33	35.19
140	135.82	33.95	135.50	35.23	135.16	36.49
145	140.67	35.17	140.33	36.49	139.99	37.80
150	145.52	36.38	145.17	37.75	144.81	39.10
155	150.37	37.59	150.01	39.00	149.64	40.40
160	155.22	38.81	154.85	40.26	154.47	41.71
165	160.07	40.02	159.69	41.52	159.30	43.01
170	164.92	41.23	164.53	42.78	164.12	44.31
175	169.77	42.44	169.37	44.04	168.95	45.62
180	174.63	43.66	174.21	45.29	173.78	46.92
185	179.48	44.87	179.05	46.55	178.60	48.22
190	184.33	46.08	183.89	47.81	183.43	49.53
195	189.18	47.29	188.73	49.07	188.26	50.83
200	194.03	48.51	193.56	50.33	193.09	52.13
205	198.88	49.72	198.40	51.58	197.91	53.44
210	203.73	50.93	203.24	52.84	202.74	54.74
215	208.58	52.15	208.08	54.10	207.57	56.04
220	213.43	53.36	212.92	55.36	212.39	57.35
225	218.28	54.57	217.76	56.62	217.22	58.65
230	223.13	55.78	222.60	57.88	222.05	59.95
235	227.98	57.00	227.44	59.13	226.88	61.26
240	232.83	58.21	232.28	60.39	231.70	62.56
245	237.68	59.42	237.12	61.65	236.53	63.86
250	242.54	60.63	241.96	62.91	241.36	65.17
255	247.39	61.85	246.79	64.17	246.18	66.47
260	252.24	63.06	251.63	65.42	251.01	67.77
265	257.09	64.27	256.47	66.68	255.84	69.08
270	261.94	65.48	261.31	67.94	260.67	70.38
275	266.79	66.70	266.15	69.20	265.49	71.68
280	271.64	67.91	270.99	70.46	270.32	72.99
285	276.49	69.12	275.83	71.72	275.15	74.29
290	281.34	70.34	280.67	72.97	279.97	75.59
295	286.19	71.55	285.51	74.23	284.80	76.90
300	291.04	72.76	290.35	75.49	289.63	78.20

SLOPE REDUCTION TABLE FOR THE PERCENT ABNEY

	28%(15° 39')		29%(16° 10')		30%(16° 42')	
SD	HD	DE	HD	DE	HD	DE
35	33.70	9.44	33.62	9.75	33.52	10.06
40	38.52	10.79	38.42	11.14	38.31	11.49
45	43.33	12.13	43.22	12.53	43.10	12.93
50	48.15	13.48	48.02	13.93	47.89	14.37
55	52.96	14.83	52.82	15.32	52.68	15.80
60	57.78	16.18	57.63	16.71	57.47	17.24
65	62.59	17.53	62.43	18.10	62.26	18.68
70	67.41	18.87	67.23	19.50	67.05	20.11
75	72.22	20.22	72.03	20.89	71.84	21.55
80	77.04	21.57	76.83	22.28	76.63	22.99
85	81.85	22.92	81.64	23.67	81.42	24.42
90	86.67	24.27	86.44	25.07	86.20	25.86
95	91.48	25.61	91.24	26.46	90.99	27.30
100	96.30	26.96	96.04	27.85	95.78	28.73
105	101.11	28.31	100.85	29.25	100.57	30.17
110	105.93	29.66	105.65	30.64	105.36	31.61
115	110.74	31.01	110.45	32.03	110.15	33.05
120	115.56	32.36	115.25	33.42	114.94	34.48
125	120.37	33.70	120.05	34.82	119.73	35.92
130	125.19	35.05	124.86	36.21	124.52	37.36
135	130.00	36.40	129.66	37.60	129.31	38.79
140	134.81	37.75	134.46	38.99	134.10	40.23
145	139.63	39.10	139.26	40.39	138.88	41.67
150	144.44	40.44	144.06	41.78	143.67	43.10
155	149.26	41.79	148.87	43.17	148.46	44.54
160	154.07	43.14	153.67	44.56	153.25	45.98
165	158.89	44.49	158.47	45.96	158.04	47.41
170	163.70	45.84	163.27	47.35	162.83	48.85
175	168.52	47.19	168.08	48.74	167.62	50.29
180	173.33	48.53	172.88	50.13	172.41	51.72
185	178.15	49.88	177.68	51.53	177.20	53.16
190	182.96	51.23	182.48	52.92	181.99	54.60
195	187.78	52.58	187.28	54.31	186.78	56.03
200	192.59	53.93	192.09	55.70	191.57	57.47
205	197.41	55.27	196.89	57.10	196.35	58.91
210	202.22	56.62	201.69	58.49	201.14	60.34
215	207.04	57.97	206.49	59.88	205.93	61.78
220	211.85	59.32	211.29	61.28	210.72	63.22
225	216.67	60.67	216.10	62.67	215.51	64.65
230	221.48	62.01	220.90	64.06	220.30	66.09
235	226.30	63.36	225.70	65.45	225.09	67.53
240	231.11	64.71	230.50	66.85	229.88	68.96
245	235.93	66.06	235.31	68.24	234.67	70.40
250	240.74	67.41	240.11	69.63	239.46	71.84
255	245.56	68.76	244.91	71.02	244.25	73.27
260	250.37	70.10	249.71	72.42	249.03	74.71
265	255.19	71.45	254.51	73.81	253.82	76.15
270	260.00	72.80	259.32	75.20	258.61	77.58
275	264.82	74.15	264.12	76.59	263.40	79.02
280	269.63	75.50	268.92	77.99	268.19	80.46
285	274.44	76.84	273.72	79.38	272.98	81.89
290	279.26	78.19	278.52	80.77	277.77	83.33
295	284.07	79.54	283.33	82.16	282.56	84.77
300	288.89	80.89	288.13	83.56	287.35	86.20

SLOPE REDUCTION TABLE FOR THE PERCENT ABNEY

	31%(17°13′)		32%(17°45′)		33%(18°16′)	
SD	HD	DE	HD	DE	HD	DE
35	33.43	10.36	33.33	10.67	33.24	10.97
40	38.21	11.84	38.10	12.19	37.99	12.54
45	42.98	13.32	42.86	13.71	42.73	14.10
50	47.76	14.80	47.62	15.24	47.48	15.67
55	52.53	16.29	52.38	16.76	52.23	17.24
60	57.31	17.77	57.15	18.29	56.98	18.80
65	62.09	19.25	61.91	19.81	61.73	20.37
70	66.86	20.73	66.67	21.33	66.47	21.94
75	71.64	22.21	71.43	22.86	71.22	23.50
80	76.41	23.69	76.19	24.38	75.97	25.07
85	81.19	25.17	80.96	25.91	80.72	26.64
90	85.96	26.65	85.72	27.43	85.47	28.20
95	90.74	28.13	90.48	28.95	90.21	29.77
100	95.52	29.61	95.24	30.48	94.96	31.34
105	100.29	31.09	100.00	32.00	99.71	32.90
110	105.07	32.57	104.77	33.53	104.46	34.47
115	109.84	34.05	109.53	35.05	109.21	36.04
120	114.62	35.53	114.29	36.57	113.96	37.61
125	119.39	37.01	119.05	38.10	118.70	39.17
130	124.17	38.49	123.82	39.62	123.45	40.74
135	128.95	39.97	128.58	41.14	128.20	42.31
140	133.72	41.45	133.34	42.67	132.95	43.87
145	138.50	42.93	138.10	44.19	137.70	45.44
150	143.27	44.41	142.86	45.72	142.44	47.01
155	148.05	45.90	147.63	47.24	147.19	48.57
160	152.83	47.38	152.39	48.76	151.94	50.14
165	157.60	48.86	157.15	50.29	156.69	51.71
170	162.38	50.34	161.91	51.81	161.44	53.27
175	167.15	51.82	166.67	53.34	166.19	54.84
180	171.93	53.30	171.44	54.86	170.93	56.41
185	176.70	54.78	176.20	56.38	175.68	57.97
190	181.48	56.26	180.96	57.91	180.43	59.54
195	186.26	57.74	185.72	59.43	185.18	61.11
200	191.03	59.22	190.48	60.96	189.93	62.68
205	195.81	60.70	195.25	62.48	194.67	64.24
210	200.58	62.18	200.01	64.00	199.42	65.81
215	205.36	63.66	204.77	65.53	204.17	67.38
220	210.13	65.14	209.53	67.05	208.92	68.94
225	214.91	66.62	214.30	68.57	213.67	70.51
230	219.69	68.10	219.06	70.10	218.41	72.08
235	224.46	69.58	223.82	71.62	223.16	73.64
240	229.24	71.06	228.58	73.15	227.91	75.21
245	234.01	72.54	233.34	74.67	232.66	76.78
250	238.79	74.02	238.11	76.19	237.41	78.34
255	243.57	75.51	242.87	77.72	242.16	79.91
260	248.34	76.99	247.63	79.24	246.90	81.48
265	253.12	78.47	252.39	80.77	251.65	83.05
270	257.89	79.95	257.15	82.29	256.40	84.61
275	262.67	81.43	261.92	83.81	261.15	86.18
280	267.44	82.91	266.68	85.34	265.90	87.75
285	272.22	84.39	271.44	86.86	270.64	89.31
290	277.00	85.87	276.20	88.38	275.39	90.88
295	281.77	87.35	280.97	89.91	280.14	92.45
300	286.55	88.83	285.73	91.43	284.89	94.01

SLOPE REDUCTION TABLE FOR THE PERCENT ABNEY

SD	34%(18°47')		35%(19°17')		36%(19°48')	
	HD	DE	HD	DE	HD	DE
35	33.14	11.27	33.04	11.56	32.93	11.86
40	37.87	12.88	37.75	13.21	37.64	13.55
45	42.60	14.49	42.47	14.87	42.34	15.24
50	47.34	16.10	47.19	16.52	47.04	16.94
55	52.07	17.70	51.91	18.17	51.75	18.63
60	56.81	19.31	56.63	19.82	56.45	20.32
65	61.54	20.92	61.35	21.47	61.16	22.02
70	66.27	22.53	66.07	23.12	65.86	23.71
75	71.01	24.14	70.79	24.78	70.57	25.40
80	75.74	25.75	75.51	26.43	75.27	27.10
85	80.48	27.36	80.23	28.08	79.98	28.79
90	85.21	28.97	84.95	29.73	84.68	30.48
95	89.94	30.58	89.67	31.38	89.38	32.18
100	94.68	32.19	94.39	33.04	94.09	33.87
05	99.41	33.80	99.11	34.69	98.79	35.57
10	104.15	35.41	103.82	36.34	103.50	37.26
15	108.88	37.02	108.54	37.99	108.20	38.95
20	113.61	38.63	113.26	39.64	112.91	40.65
25	118.35	40.24	117.98	41.29	117.61	42.34
30	123.08	41.85	122.70	42.95	122.32	44.03
35	127.81	43.46	127.42	44.60	127.02	45.73
40	132.55	45.07	132.14	46.25	131.72	47.42
45	137.28	46.68	136.86	47.90	136.43	49.11
150	142.02	48.29	141.58	49.55	141.13	50.81
155	146.75	49.89	146.30	51.20	145.84	52.50
160	151.48	51.50	151.02	52.86	150.54	54.20
165	156.22	53.11	155.74	54.51	155.25	55.89
170	160.95	54.72	160.46	56.16	159.95	57.58
175	165.69	56.33	165.18	57.81	164.66	59.28
180	170.42	57.94	169.89	59.46	169.36	60.97
185	175.15	59.55	174.61	61.11	174.06	62.66
190	179.89	61.16	179.33	62.77	178.77	64.36
195	184.62	62.77	184.05	64.42	183.47	66.05
200	189.35	64.38	188.77	66.07	188.18	67.74
205	194.09	65.99	193.49	67.72	192.88	69.44
210	198.82	67.60	198.21	69.37	197.59	71.13
215	203.56	69.21	202.93	71.03	202.29	72.82
220	208.29	70.82	207.65	72.68	207.00	74.52
225	213.02	72.43	212.37	74.33	211.70	76.21
230	217.76	74.04	217.09	75.98	216.40	77.91
235	222.49	75.65	221.81	77.63	221.11	79.60
240	227.23	77.26	226.53	79.28	225.81	81.29
245	231.96	78.87	231.25	80.94	230.52	82.99
250	236.69	80.48	235.96	82.59	235.22	84.68
255	241.43	82.09	240.68	84.24	239.93	86.37
260	246.16	83.69	245.40	85.89	244.63	88.07
265	250.89	85.30	250.12	87.54	249.34	89.76
270	255.63	86.91	254.84	89.19	254.04	91.45
275	260.36	88.52	259.56	90.85	258.74	93.15
280	265.10	90.13	264.28	92.50	263.45	94.84
285	269.83	91.74	269.00	94.15	268.15	96.54
290	274.56	93.35	273.72	95.80	272.86	98.23
295	279.30	94.96	278.44	97.45	277.56	99.92
300	284.03	96.57	283.16	99.11	282.27	101.62

SLOPE REDUCTION TABLE FOR THE PERCENT ABNEY

SD	37%(20°18′) HD	DE	38%(20°48′) HD	DE	39%(21°18′) HD	DE
35	32.83	12.15	32.72	12.43	32.61	12.72
40	37.51	13.88	37.39	14.21	37.27	14.53
45	42.20	15.62	42.07	15.98	41.92	16.35
50	46.89	17.35	46.74	17.76	46.58	18.17
55	51.58	19.09	51.41	19.54	51.24	19.98
60	56.27	20.82	56.09	21.31	55.90	21.80
65	60.96	22.56	60.76	23.09	60.56	23.62
70	65.65	24.29	65.43	24.87	65.22	25.43
75	70.34	26.03	70.11	26.64	69.87	27.25
80	75.03	27.76	74.78	28.42	74.53	29.07
85	79.72	29.50	79.46	30.19	79.19	30.88
90	84.41	31.23	84.13	31.97	83.85	32.70
95	89.10	32.97	88.80	33.75	88.51	34.52
100	93.79	34.70	93.48	35.52	93.17	36.33
105	98.48	36.44	98.15	37.30	97.82	38.15
110	103.16	38.17	102.83	39.07	102.48	39.97
115	107.85	39.91	107.50	40.85	107.14	41.78
120	112.54	41.64	112.17	42.63	111.80	43.60
125	117.23	43.38	116.85	44.40	116.46	45.42
130	121.92	45.11	121.52	46.18	121.12	47.23
135	126.61	46.85	126.20	47.95	125.77	49.05
140	131.30	48.58	130.87	49.73	130.43	50.87
145	135.99	50.32	135.54	51.51	135.09	52.69
150	140.68	52.05	140.22	53.28	139.75	54.50
155	145.37	53.79	144.89	55.06	144.41	56.32
160	150.06	55.52	149.57	56.83	149.06	58.14
165	154.75	57.26	154.24	58.61	153.72	59.95
170	159.44	58.99	158.91	60.39	158.38	61.77
175	164.13	60.73	163.59	62.16	163.04	63.59
180	168.82	62.46	168.26	63.94	167.70	65.40
185	173.50	64.20	172.93	65.72	172.36	67.22
190	178.19	65.93	177.61	67.49	177.01	69.04
195	182.88	67.67	182.28	69.27	181.67	70.85
200	187.57	69.40	186.96	71.04	186.33	72.67
205	192.26	71.14	191.63	72.82	190.99	74.49
210	196.95	72.87	196.30	74.60	195.65	76.30
215	201.64	74.61	200.98	76.37	200.31	78.12
220	206.33	76.34	205.65	78.15	204.96	79.94
225	211.02	78.08	210.33	79.92	209.62	81.75
230	215.71	79.81	215.00	81.70	214.28	83.57
235	220.40	81.55	219.67	83.48	218.94	85.39
240	225.09	83.28	224.35	85.25	223.60	87.20
245	229.78	85.02	229.02	87.03	228.26	89.02
250	234.47	86.75	233.70	88.80	232.91	90.84
255	239.15	88.49	238.37	90.58	237.57	92.65
260	243.84	90.22	243.04	92.36	242.23	94.47
265	248.53	91.96	247.72	94.13	246.89	96.29
270	253.22	93.69	252.39	95.91	251.55	98.10
275	257.91	95.43	257.07	97.68	256.21	99.92
280	262.60	97.16	261.74	99.46	260.86	101.74
285	267.29	98.90	266.41	101.24	265.52	103.55
290	271.98	100.63	271.09	103.01	270.18	105.37
295	276.67	102.37	275.76	104.79	274.84	107.19
300	281.36	104.10	280.44	106.57	279.50	109.00

SLOPE REDUCTION TABLE FOR THE PERCENT ABNEY

SD	40%(21°48′) HD	40%(21°48′) DE	41%(22°18′) HD	41%(22°18′) DE	42%(22°47′) HD	42%(22°47′) DE
35	32.50	13.00	32.38	13.28	32.27	13.55
40	37.14	14.86	37.01	15.17	36.88	15.49
45	41.78	16.71	41.64	17.07	41.49	17.43
50	46.42	18.57	46.26	18.97	46.10	19.36
55	51.07	20.43	50.89	20.86	50.71	21.30
60	55.71	22.28	55.52	22.76	55.32	23.23
65	60.35	24.14	60.14	24.66	59.93	25.17
70	64.99	26.00	64.77	26.55	64.54	27.11
75	69.64	27.85	69.39	28.45	69.15	29.04
80	74.28	29.71	74.02	30.35	73.76	30.98
85	78.92	31.57	78.65	32.25	78.37	32.91
90	83.56	33.43	83.27	34.14	82.98	34.85
95	88.21	35.28	87.90	36.04	87.59	36.79
100	92.85	37.14	92.53	37.94	92.20	38.72
105	97.49	39.00	97.15	39.83	96.81	40.66
110	102.13	40.85	101.78	41.73	101.42	42.60
115	106.77	42.71	106.40	43.63	106.03	44.53
120	111.42	44.57	111.03	45.52	110.64	46.47
125	116.06	46.42	115.66	47.42	115.25	48.40
130	120.70	48.28	120.28	49.32	119.86	50.34
135	125.34	50.14	124.91	51.21	124.47	52.28
140	129.99	51.99	129.54	53.11	129.08	54.21
145	134.63	53.85	134.16	55.01	133.69	56.15
150	139.27	55.71	138.79	56.90	138.30	58.08
155	143.91	57.57	143.41	58.80	142.91	60.02
160	148.56	59.42	148.04	60.70	147.52	61.96
165	153.20	61.28	152.67	62.59	152.13	63.89
170	157.84	63.14	157.29	64.49	156.74	65.83
175	162.48	64.99	161.92	66.39	161.35	67.77
180	167.13	66.85	166.55	68.28	165.96	69.70
185	171.77	68.71	171.17	70.18	170.57	71.64
190	176.41	70.56	175.80	72.08	175.18	73.57
195	181.05	72.42	180.42	73.97	179.79	75.51
200	185.70	74.28	185.05	75.87	184.40	77.45
205	190.34	76.14	189.68	77.77	189.01	79.38
210	194.98	77.99	194.30	79.66	193.62	81.32
215	199.62	79.85	198.93	81.56	198.23	83.25
220	204.26	81.71	203.56	83.46	202.84	85.19
225	208.91	83.56	208.18	85.35	207.45	87.13
230	213.55	85.42	212.81	87.25	212.06	89.06
235	218.19	87.28	217.43	89.15	216.67	91.00
240	222.83	89.13	222.06	91.04	221.28	92.94
245	227.48	90.99	226.69	92.94	225.89	94.87
250	232.12	92.85	231.31	94.84	230.50	96.81
255	236.76	94.70	235.94	96.74	235.11	98.74
260	241.40	96.56	240.57	98.63	239.72	100.68
265	246.05	98.42	245.19	100.53	244.33	102.62
270	250.69	100.28	249.82	102.43	248.94	104.55
275	255.33	102.13	254.44	104.32	253.55	106.49
280	259.97	103.99	259.07	106.22	258.15	108.43
285	264.62	105.85	263.70	108.12	262.76	110.36
290	269.26	107.70	268.32	110.01	267.37	112.30
295	273.90	109.56	272.95	111.91	271.98	114.23
300	278.54	111.42	277.58	113.81	276.59	116.17

SLOPE REDUCTION TABLE FOR THE PERCENT ABNEY

	43%(23° 16′)		44%(23° 45′)		45%(24° 14′)	
SD	HD	DE	HD	DE	HD	DE
35	32.15	13.83	32.04	14.10	31.92	14.36
40	36.75	15.80	36.61	16.11	36.48	16.41
45	41.34	17.78	41.19	18.12	41.04	18.47
50	45.93	19.75	45.77	20.14	45.60	20.52
55	50.53	21.73	50.34	22.15	50.16	22.57
60	55.12	23.70	54.92	24.16	54.72	24.62
65	59.71	25.68	59.50	26.18	59.27	26.67
70	64.31	27.65	64.07	28.19	63.83	28.73
75	68.90	29.63	68.65	30.21	68.39	30.78
80	73.49	31.60	73.23	32.22	72.95	32.83
85	78.09	33.58	77.80	34.23	77.51	34.88
90	82.68	35.55	82.38	36.25	82.07	36.93
95	87.27	37.53	86.95	38.26	86.63	38.98
100	91.87	39.50	91.53	40.27	91.19	41.04
105	96.46	41.48	96.11	42.29	95.75	43.09
110	101.05	43.45	100.68	44.30	100.31	45.14
115	105.65	45.43	105.26	46.31	104.87	47.19
120	110.24	47.40	109.84	48.33	109.43	49.24
125	114.83	49.38	114.41	50.34	113.99	51.30
130	119.43	51.35	118.99	52.36	118.55	53.35
135	124.02	53.33	123.57	54.37	123.11	55.40
140	128.61	55.30	128.14	56.38	127.67	57.45
145	133.21	57.28	132.72	58.40	132.23	59.50
150	137.80	59.25	137.30	60.41	136.79	61.55
155	142.39	61.23	141.87	62.42	141.35	63.61
160	146.99	63.20	146.45	64.44	145.91	65.66
165	151.58	65.18	151.03	66.45	150.47	67.71
170	156.17	67.15	155.60	68.47	155.03	69.76
175	160.77	69.13	160.18	70.48	159.59	71.81
180	165.36	71.10	164.76	72.49	164.15	73.87
185	169.95	73.08	169.33	74.51	168.71	75.92
190	174.55	75.06	173.91	76.52	173.27	77.97
195	179.14	77.03	178.49	78.53	177.82	80.02
200	183.73	79.01	183.06	80.55	182.38	82.07
205	188.33	80.98	187.64	82.56	186.94	84.12
210	192.92	82.96	192.22	84.58	191.50	86.18
215	197.51	84.93	196.79	86.59	196.06	88.23
220	202.11	86.91	201.37	88.60	200.62	90.28
225	206.70	88.88	205.95	90.62	205.18	92.33
230	211.29	90.86	210.52	92.63	209.74	94.38
235	215.89	92.83	215.10	94.64	214.30	96.44
240	220.48	94.81	219.68	96.66	218.86	98.49
245	225.07	96.78	224.25	98.67	223.42	100.54
250	229.67	98.76	228.83	100.68	227.98	102.59
255	234.26	100.73	233.41	102.70	232.54	104.64
260	238.85	102.71	237.98	104.71	237.10	106.69
265	243.45	104.68	242.56	106.73	241.66	108.75
270	248.04	106.66	247.14	108.74	246.22	110.80
275	252.63	108.63	251.71	110.75	250.78	112.85
280	257.23	110.61	256.29	112.77	255.34	114.90
285	261.82	112.58	260.86	114.78	259.90	116.95
290	266.41	114.56	265.44	116.79	264.46	119.01
295	271.01	116.53	270.02	118.81	269.02	121.06
300	275.60	118.51	274.59	120.82	273.58	123.11

SLOPE REDUCTION TABLE FOR THE PERCENT ABNEY

| | 46%(24°42′) | | 47%(25°10′) | | 48%(25°38′) | |
SO	HD	DE	HD	DE	HD	DE
35	31.80	14.63	31.68	14.89	31.55	15.15
40	36.34	16.72	36.20	17.01	36.06	17.31
45	40.88	18.81	40.73	19.14	40.57	19.47
50	45.42	20.90	45.25	21.27	45.08	21.64
55	49.97	22.98	49.78	23.39	49.58	23.80
60	54.51	25.07	54.30	25.52	54.09	25.96
65	59.05	27.16	58.83	27.65	58.60	28.13
70	63.59	29.25	63.35	29.78	63.11	30.29
75	68.14	31.34	67.88	31.90	67.61	32.45
80	72.68	33.43	72.40	34.03	72.12	34.62
85	77.22	35.52	76.93	36.16	76.63	36.78
90	81.76	37.61	81.45	38.28	81.14	38.95
95	86.31	39.70	85.98	40.41	85.64	41.11
100	90.85	41.79	90.50	42.54	90.15	43.27
105	95.39	43.88	95.03	44.66	94.66	45.44
110	99.93	45.97	99.55	46.79	99.17	47.60
115	104.48	48.06	104.08	48.92	103.68	49.76
120	109.02	50.15	108.60	51.04	108.18	51.93
125	113.56	52.24	113.13	53.17	112.69	54.09
130	118.10	54.33	117.65	55.30	117.20	56.26
135	122.65	56.42	122.18	57.42	121.71	58.42
140	127.19	58.51	126.70	59.55	126.21	60.58
145	131.73	60.60	131.23	61.68	130.72	62.75
150	136.27	62.69	135.75	63.80	135.23	64.91
155	140.82	64.78	140.28	65.93	139.74	67.07
160	145.36	66.86	144.80	68.06	144.24	69.24
165	149.90	68.95	149.33	70.18	148.75	71.40
170	154.44	71.04	153.85	72.31	153.26	73.56
175	158.99	73.13	158.38	74.44	157.77	75.73
180	163.53	75.22	162.90	76.57	162.27	77.89
185	168.07	77.31	167.43	78.69	166.78	80.06
190	172.61	79.40	171.95	80.82	171.29	82.22
195	177.16	81.49	176.48	82.95	175.80	84.38
200	181.70	83.58	181.00	85.07	180.30	86.55
205	186.24	85.67	185.53	87.20	184.81	88.71
210	190.78	87.76	190.05	89.33	189.32	90.87
215	195.33	89.85	194.58	91.45	193.83	93.04
220	199.87	91.94	199.11	93.58	198.34	95.20
225	204.41	94.03	203.63	95.71	202.84	97.36
230	208.95	96.12	208.16	97.83	207.35	99.53
235	213.50	98.21	212.68	99.96	211.86	101.69
240	218.04	100.30	217.21	102.09	216.37	103.86
245	222.58	102.39	221.73	104.21	220.87	106.02
250	227.12	104.48	226.26	106.34	225.38	108.18
255	231.67	106.57	230.78	108.47	229.89	110.35
260	236.21	108.66	235.31	110.59	234.40	112.51
265	240.75	110.74	239.83	112.72	238.90	114.67
270	245.29	112.83	244.36	114.85	243.41	116.84
275	249.83	114.92	248.88	116.97	247.92	119.00
280	254.38	117.01	253.41	119.10	252.43	121.16
285	258.92	119.10	257.93	121.23	256.93	123.33
290	263.46	121.19	262.46	123.35	261.44	125.49
295	268.00	123.28	266.98	125.48	265.95	127.66
300	272.55	125.37	271.51	127.61	270.46	129.82

SLOPE REDUCTION TABLE FOR THE PERCENT ABNEY

SD	49%(26° 06′)		50%(26° 34′)		51%(27° 01′)	
	HD	DE	HD	DE	HD	DE
35	31.43	15.40	31.30	15.65	31.18	15.90
40	35.92	17.60	35.78	17.89	35.63	18.17
45	40.41	19.80	40.25	20.12	40.09	20.44
50	44.90	22.00	44.72	22.36	44.54	22.72
55	49.39	24.20	49.19	24.60	49.00	24.99
60	53.88	26.40	53.67	26.83	53.45	27.26
65	58.37	28.60	58.14	29.07	57.90	29.53
70	62.86	30.80	62.61	31.30	62.36	31.80
75	67.35	33.00	67.08	33.54	66.81	34.07
80	71.84	35.20	71.55	35.78	71.27	36.35
85	76.33	37.40	76.03	38.01	75.72	38.62
90	80.82	39.60	80.50	40.25	80.18	40.89
95	85.31	41.80	84.97	42.49	84.63	43.16
100	89.80	44.00	89.44	44.72	89.08	45.43
105	94.29	46.20	93.91	46.96	93.54	47.70
110	98.78	48.40	98.39	49.19	97.99	49.98
115	103.27	50.60	102.86	51.43	102.45	52.25
120	107.76	52.80	107.33	53.67	106.90	54.52
125	112.25	55.00	111.80	55.90	111.35	56.79
130	116.74	57.20	116.28	58.14	115.81	59.06
135	121.23	59.40	120.75	60.37	120.26	61.33
140	125.72	61.60	125.22	62.61	124.72	63.61
145	130.21	63.80	129.69	64.85	129.17	65.88
150	134.70	66.00	134.16	67.08	133.63	68.15
155	139.19	68.20	138.64	69.32	138.08	70.42
160	143.68	70.40	143.11	71.55	142.53	72.69
165	148.17	72.60	147.58	73.79	146.99	74.96
170	152.66	74.80	152.05	76.03	151.44	77.24
175	157.15	77.00	156.52	78.26	155.90	79.51
180	161.64	79.20	161.00	80.50	160.35	81.78
185	166.13	81.40	165.47	82.73	164.80	84.05
190	170.62	83.60	169.94	84.97	169.26	86.32
195	175.11	85.80	174.41	87.21	173.71	88.59
200	179.60	88.00	178.89	89.44	178.17	90.87
205	184.09	90.20	183.36	91.68	182.62	93.14
210	188.58	92.40	187.83	93.91	187.08	95.41
215	193.07	94.60	192.30	96.15	191.53	97.68
220	197.56	96.80	196.77	98.39	195.98	99.95
225	202.05	99.00	201.25	100.62	200.44	102.22
230	206.54	101.20	205.72	102.86	204.89	104.49
235	211.03	103.40	210.19	105.10	209.35	106.77
240	215.52	105.60	214.66	107.33	213.80	109.04
245	220.01	107.80	219.13	109.57	218.25	111.31
250	224.50	110.00	223.61	111.80	222.71	113.58
255	228.99	112.20	228.08	114.04	227.16	115.85
260	233.48	114.40	232.55	116.28	231.62	118.12
265	237.97	116.60	237.02	118.51	236.07	120.40
270	242.46	118.80	241.50	120.75	240.53	122.67
275	246.95	121.00	245.97	122.98	244.98	124.94
280	251.44	123.20	250.44	125.22	249.43	127.21
285	255.93	125.40	254.91	127.46	253.89	129.48
290	260.42	127.60	259.38	129.69	258.34	131.75
295	264.91	129.80	263.86	131.93	262.80	134.03
300	269.40	132.00	268.33	134.16	267.25	136.30

SLOPE REDUCTION TABLE FOR THE PERCENT ABNEY

SD	52% (27° 28')		53% (27° 55')		54% (28° 22')	
	HD	DE	HD	DE	HD	DE
35	31.05	16.15	30.93	16.39	30.80	16.63
40	35.49	18.45	35.34	18.73	35.20	19.01
45	39.92	20.76	39.76	21.07	39.60	21.38
50	44.36	23.07	44.18	23.41	44.00	23.76
55	48.80	25.37	48.60	25.76	48.39	26.13
60	53.23	27.68	53.01	28.10	52.79	28.51
65	57.67	29.99	57.43	30.44	57.19	30.88
70	62.11	32.29	61.85	32.78	61.59	33.26
75	66.54	34.60	66.27	35.12	65.99	35.64
80	70.98	36.91	70.69	37.46	70.39	38.01
85	75.41	39.21	75.10	39.80	74.79	40.39
90	79.85	41.52	79.52	42.15	79.19	42.76
95	84.29	43.83	83.94	44.49	83.59	45.14
100	88.72	46.14	88.36	46.83	87.99	47.51
105	93.16	48.44	92.78	49.17	92.39	49.89
110	97.59	50.75	97.19	51.51	96.79	52.27
115	102.03	53.06	101.61	53.85	101.19	54.64
120	106.47	55.36	106.03	56.20	105.59	57.02
125	110.90	57.67	110.45	58.54	109.99	59.39
130	115.34	59.98	114.86	60.88	114.39	61.77
135	119.77	62.28	119.28	63.22	118.79	64.15
140	124.21	64.59	123.70	65.56	123.19	66.52
145	128.65	66.90	128.12	67.90	127.59	68.90
150	133.08	69.20	132.54	70.24	131.99	71.27
155	137.52	71.51	136.95	72.59	136.39	73.65
160	141.95	73.82	141.37	74.93	140.78	76.02
165	146.39	76.12	145.79	77.27	145.18	78.40
170	150.83	78.43	150.21	79.61	149.58	80.78
175	155.26	80.74	154.63	81.95	153.98	83.15
180	159.70	83.04	159.04	84.29	158.38	85.53
185	164.14	85.35	163.46	86.63	162.78	87.90
190	168.57	87.66	167.88	88.98	167.18	90.28
195	173.01	89.96	172.30	91.32	171.58	92.65
200	177.44	92.27	176.71	93.66	175.98	95.03
205	181.88	94.58	181.13	96.00	180.38	97.41
210	186.32	96.88	185.55	98.34	184.78	99.78
215	190.75	99.19	189.97	100.68	189.18	102.16
220	195.19	101.50	194.39	103.02	193.58	104.53
225	199.62	103.80	198.80	105.37	197.98	106.91
230	204.06	106.11	203.22	107.71	202.38	109.28
235	208.50	108.42	207.64	110.05	206.78	111.66
240	212.93	110.72	212.06	112.39	211.18	114.04
245	217.37	113.03	216.48	114.73	215.58	116.41
250	221.80	115.34	220.89	117.07	219.98	118.79
255	226.24	117.64	225.31	119.41	224.38	121.16
260	230.68	119.95	229.73	121.76	228.78	123.54
265	235.11	122.26	234.15	124.10	233.17	125.91
270	239.55	124.57	238.56	126.44	237.57	128.29
275	243.98	126.87	242.98	128.78	241.97	130.67
280	248.42	129.18	247.40	131.12	246.37	133.04
285	252.86	131.49	251.82	133.46	250.77	135.42
290	257.29	133.79	256.24	135.81	255.17	137.79
295	261.73	136.10	260.65	138.15	259.57	140.17
300	266.17	138.41	265.07	140.49	263.97	142.54

SLOPE REDUCTION TABLE FOR THE PERCENT ABNEY

SD	55%(28° 49') HD	DE	56%(29° 15') HD	DE	57%(29° 41') HD	DE
35	30.67	16.87	30.54	17.10	30.41	17.33
40	35.05	19.28	34.90	19.54	34.75	19.81
45	39.43	21.69	39.26	21.99	39.09	22.28
50	43.81	24.10	43.63	24.43	43.44	24.76
55	48.19	26.51	47.99	26.87	47.78	27.24
60	52.57	28.92	52.35	29.32	52.13	29.71
65	56.95	31.32	56.71	31.76	56.47	32.19
70	61.34	33.73	61.08	34.20	60.81	34.66
75	65.72	36.14	65.44	36.65	65.16	37.14
80	70.10	38.55	69.80	39.09	69.50	39.62
85	74.48	40.96	74.16	41.53	73.85	42.09
90	78.86	43.37	78.53	43.97	78.19	44.57
95	83.24	45.78	82.89	46.42	82.53	47.04
100	87.62	48.19	87.25	48.86	86.88	49.52
105	92.00	50.60	91.61	51.30	91.22	52.00
110	96.38	53.01	95.98	53.75	95.57	54.47
115	100.76	55.42	100.34	56.19	99.91	56.95
120	105.15	57.83	104.70	58.63	104.25	59.42
125	109.53	60.24	109.06	61.08	108.60	61.90
130	113.91	62.65	113.43	63.52	112.94	64.38
135	118.29	65.06	117.79	65.96	117.28	66.85
140	122.67	67.47	122.15	68.40	121.63	69.33
145	127.05	69.88	126.51	70.85	125.97	71.80
150	131.43	72.29	130.88	73.29	130.32	74.28
155	135.81	74.70	135.24	75.73	134.66	76.76
160	140.19	77.11	139.60	78.18	139.00	79.23
165	144.58	79.52	143.96	80.62	143.35	81.71
170	148.96	81.93	148.33	83.06	147.69	84.18
175	153.34	84.34	152.69	85.51	152.04	86.66
180	157.72	86.75	157.05	87.95	156.38	89.14
185	162.10	89.15	161.41	90.39	160.72	91.61
190	166.48	91.56	165.78	92.83	165.07	94.09
195	170.86	93.97	170.14	95.28	169.41	96.56
200	175.24	96.38	174.50	97.72	173.76	99.04
205	179.62	98.79	178.86	100.16	178.10	101.52
210	184.01	101.20	183.23	102.61	182.44	103.99
215	188.39	103.61	187.59	105.05	186.79	106.47
220	192.77	106.02	191.95	107.49	191.13	108.94
225	197.15	108.43	196.31	109.94	195.47	111.42
230	201.53	110.84	200.68	112.38	199.82	113.90
235	205.91	113.25	205.04	114.82	204.16	116.37
240	210.29	115.66	209.40	117.26	208.51	118.85
245	214.67	118.07	213.76	119.71	212.85	121.32
250	219.05	120.48	218.13	122.15	217.19	123.80
255	223.44	122.89	222.49	124.59	221.54	126.28
260	227.82	125.30	226.85	127.04	225.88	128.75
265	232.20	127.71	231.21	129.48	230.23	131.23
270	236.58	130.12	235.58	131.92	234.57	133.70
275	240.96	132.53	239.94	134.37	238.91	136.18
280	245.34	134.94	244.30	136.81	243.26	138.66
285	249.72	137.35	248.66	139.25	247.60	141.13
290	254.10	139.76	253.03	141.69	251.95	143.61
295	258.48	142.17	257.39	144.14	256.29	146.08
300	262.86	144.58	261.75	146.58	260.63	148.56

SLOPE REDUCTION TABLE FOR THE PERCENT ABNEY

	58%(30° 07′)		59%(30° 32′)		60%(30° 58′)	
SD	HD	DE	HD	DE	HD	DE
35	30.28	17.56	30.14	17.79	30.01	18.01
40	34.60	20.07	34.45	20.33	34.30	20.58
45	38.93	22.58	38.76	22.87	38.59	23.15
50	43.25	25.09	43.06	25.41	42.87	25.72
55	47.58	27.59	47.37	27.95	47.16	28.30
60	51.90	30.10	51.68	30.49	51.45	30.87
65	56.23	32.61	55.98	33.03	55.74	33.44
70	60.55	35.12	60.29	35.57	60.02	36.01
75	64.88	37.63	64.60	38.11	64.31	38.59
80	69.20	40.14	68.90	40.65	68.60	41.16
85	73.53	42.65	73.21	43.19	72.89	43.73
90	77.85	45.15	77.51	45.73	77.17	46.30
95	82.18	47.66	81.82	48.27	81.46	48.88
100	86.50	50.17	86.13	50.81	85.75	51.45
105	90.83	52.68	90.43	53.36	90.04	54.02
110	95.15	55.19	94.74	55.90	94.32	56.59
115	99.48	57.70	99.05	58.44	98.61	59.17
120	103.80	60.21	103.35	60.98	102.90	61.74
125	108.13	62.71	107.66	63.52	107.19	64.31
130	112.45	65.22	111.97	66.06	111.47	66.88
135	116.78	67.73	116.27	68.60	115.76	69.46
140	121.10	70.24	120.58	71.14	120.05	72.03
145	125.43	72.75	124.88	73.68	124.34	74.60
150	129.75	75.26	129.19	76.22	128.62	77.17
155	134.08	77.77	133.50	78.76	132.91	79.75
160	138.40	80.27	137.80	81.30	137.20	82.32
165	142.73	82.78	142.11	83.84	141.49	84.89
170	147.06	85.29	146.42	86.39	145.77	87.46
175	151.38	87.80	150.72	88.93	150.06	90.04
180	155.71	90.31	155.03	91.47	154.35	92.61
185	160.03	92.82	159.33	94.01	158.64	95.18
190	164.36	95.33	163.64	96.55	162.92	97.75
195	168.68	97.84	167.95	99.09	167.21	100.33
200	173.01	100.34	172.25	101.63	171.50	102.90
205	177.33	102.85	176.56	104.17	175.79	105.47
210	181.66	105.36	180.87	106.71	180.07	108.04
215	185.98	107.87	185.17	109.25	184.36	110.62
220	190.31	110.38	189.43	111.79	188.65	113.19
225	194.63	112.89	193.79	114.33	192.94	115.76
230	198.96	115.40	198.09	116.87	197.22	118.33
235	203.28	117.90	202.40	119.41	201.51	120.91
240	207.61	120.41	206.70	121.96	205.80	123.48
245	211.93	122.92	211.01	124.50	210.09	126.05
250	216.26	125.43	215.32	127.04	214.37	128.62
255	220.58	127.94	219.62	129.58	218.66	131.20
260	224.91	130.45	223.93	132.12	222.95	133.77
265	229.23	132.96	228.24	134.66	227.24	136.34
270	233.56	135.46	232.54	137.20	231.52	138.91
275	237.88	137.97	236.85	139.74	235.81	141.49
280	242.21	140.48	241.16	142.28	240.10	144.06
285	246.53	142.99	245.46	144.82	244.39	146.63
290	250.86	145.50	249.77	147.36	248.67	149.20
295	255.18	148.01	254.07	149.90	252.96	151.78
300	259.51	150.52	258.38	152.44	257.25	154.35

SLOPE REDUCTION TABLE FOR THE PERCENT ABNEY

SD	61%(31° 23′) HD	OE	62%(31° 48′) HD	OE	63%(32° 13′) HD	OE
35	29.88	18.23	29.75	18.44	29.61	18.66
40	34.15	20.83	34.00	21.08	33.84	21.32
45	38.42	23.43	38.25	23.71	38.07	23.99
50	42.69	26.04	42.50	26.35	42.30	26.65
55	46.95	28.64	46.74	28.98	46.54	29.32
60	51.22	31.25	50.99	31.62	50.77	31.98
65	55.49	33.85	55.24	34.25	55.00	34.65
70	59.76	36.45	59.49	36.89	59.23	37.31
75	64.03	39.06	63.74	39.52	63.46	39.98
80	68.30	41.66	67.99	42.16	67.69	42.64
85	72.56	44.26	72.24	44.79	71.92	45.31
90	76.83	46.87	76.49	47.42	76.15	47.97
95	81.10	49.47	80.74	50.06	80.38	50.64
100	85.37	52.08	84.99	52.69	84.61	53.30
105	89.64	54.68	89.24	55.33	88.84	55.97
110	93.91	57.28	93.49	57.96	93.07	58.63
115	98.18	59.89	97.74	60.60	97.30	61.30
120	102.44	62.49	101.99	63.23	101.53	63.96
125	106.71	65.09	106.24	65.87	105.76	66.63
130	110.98	67.70	110.49	68.50	109.99	69.29
135	115.25	70.30	114.74	71.14	114.22	71.96
140	119.52	72.91	118.99	73.77	118.45	74.63
145	123.79	75.51	123.24	76.41	122.68	77.29
150	128.06	78.11	127.49	79.04	126.91	79.96
155	132.32	80.72	131.73	81.68	131.14	82.62
160	136.59	83.32	135.98	84.31	135.37	85.29
165	140.86	85.93	140.23	86.95	139.61	87.95
170	145.13	88.53	144.48	89.58	143.84	90.62
175	149.40	91.13	148.73	92.21	148.07	93.28
180	153.67	93.74	152.98	94.85	152.30	95.95
185	157.94	96.34	157.23	97.48	156.53	98.61
190	162.20	98.94	161.48	100.12	160.76	101.28
195	166.47	101.55	165.73	102.75	164.99	103.94
200	170.74	104.15	169.98	105.39	169.22	106.61
205	175.01	106.76	174.23	108.02	173.45	109.27
210	179.28	109.36	178.48	110.66	177.68	111.94
215	183.55	111.96	182.73	113.29	181.91	114.60
220	187.81	114.57	186.98	115.93	186.14	117.27
225	192.08	117.17	191.23	118.56	190.37	119.93
230	196.35	119.77	195.48	121.20	194.60	122.60
235	200.62	122.38	199.73	123.83	198.83	125.26
240	204.89	124.98	203.98	126.47	203.06	127.93
245	209.16	127.59	208.23	129.10	207.29	130.59
250	213.43	130.19	212.48	131.73	211.52	133.26
255	217.69	132.79	216.73	134.37	215.75	135.92
260	221.96	135.40	220.97	137.00	219.98	138.59
265	226.23	138.00	225.22	139.64	224.21	141.25
270	230.50	140.50	229.47	142.27	228.44	143.92
275	234.77	143.21	233.72	144.91	232.68	146.59
280	239.04	145.81	237.97	147.54	236.91	149.25
285	243.31	148.42	242.22	150.18	241.14	151.92
290	247.57	151.02	246.47	152.81	245.37	154.58
295	251.84	153.62	250.72	155.45	249.60	157.25
300	256.11	156.23	254.97	158.08	253.83	159.91

SLOPE REDUCTION TABLE FOR THE PERCENT ABNEY

SD	64%(32° 37')		65%(33° 01')		66%(33° 25')	
	HD	DE	HD	DE	HD	DE
35	29.48	18.87	29.35	19.07	29.21	19.28
40	33.69	21.56	33.54	21.80	33.38	22.03
45	37.90	24.26	37.73	24.52	37.56	24.79
50	42.11	26.95	41.92	27.25	41.73	27.54
55	46.32	29.65	46.11	29.97	45.90	30.30
60	50.54	32.34	50.31	32.70	50.08	33.05
65	54.75	35.04	54.50	35.42	54.25	35.80
70	58.96	37.73	58.69	38.15	58.42	38.56
75	63.17	40.43	62.88	40.87	62.60	41.31
80	67.38	43.12	67.08	43.50	66.77	44.07
85	71.59	45.32	71.27	46.32	70.94	46.82
90	75.80	48.51	75.46	49.05	75.11	49.58
95	80.02	51.21	79.65	51.77	79.29	52.33
100	84.23	53.91	83.84	54.50	83.46	55.08
105	88.44	56.60	88.04	57.22	87.63	57.84
110	92.65	59.30	92.23	59.95	91.81	60.59
115	96.86	61.99	96.42	62.67	95.98	63.35
120	101.07	64.69	100.61	65.40	100.15	66.10
125	105.28	67.38	104.61	68.12	104.33	68.86
130	109.50	70.08	109.00	70.85	108.50	71.61
135	113.71	72.77	113.19	73.57	112.67	74.36
140	117.92	75.47	117.38	76.30	116.85	77.12
145	122.13	78.16	121.57	79.02	121.02	79.87
150	126.34	80.86	125.77	81.75	125.19	82.63
155	130.55	83.55	129.96	84.47	129.36	85.38
160	134.76	86.25	134.15	87.20	133.54	88.13
165	138.97	89.34	138.34	89.92	137.71	90.89
170	143.19	91.64	142.54	92.65	141.88	93.64
175	147.40	94.33	146.73	95.37	146.06	96.40
180	151.61	97.03	150.92	98.10	150.23	99.15
185	155.82	99.72	155.11	100.82	154.40	101.91
190	160.03	102.42	159.30	103.55	158.58	104.66
195	164.24	105.12	163.50	106.27	162.75	107.41
200	168.45	107.81	167.69	109.00	166.92	110.17
205	172.67	110.51	171.38	111.72	171.09	112.92
210	176.88	113.20	176.07	114.45	175.27	115.68
215	181.09	115.90	180.27	117.17	179.44	118.43
220	185.30	118.59	184.46	119.90	183.61	121.19
225	189.51	121.29	188.65	122.62	187.79	123.94
230	193.72	123.98	192.84	125.35	191.96	126.69
235	197.93	126.68	197.03	128.07	196.13	129.45
240	202.15	129.37	201.23	130.80	200.31	132.20
245	206.36	132.07	205.42	133.52	204.48	134.96
250	210.57	134.76	209.61	136.25	208.65	137.71
255	214.78	137.46	213.80	138.97	212.83	140.46
260	218.99	140.15	218.00	141.70	217.00	143.22
265	223.20	142.85	222.19	144.42	221.17	145.97
270	227.41	145.54	226.38	147.15	225.34	148.73
275	231.63	148.24	230.57	149.87	229.52	151.48
280	235.84	150.94	234.76	152.59	233.69	154.24
285	240.05	153.63	238.96	155.32	237.86	156.99
290	244.26	156.33	243.15	158.05	242.04	159.74
295	248.47	159.02	247.34	160.77	246.21	162.50
300	252.68	161.72	251.53	163.50	250.38	165.25

SLOPE REDUCTION TABLE FOR THE PERCENT ABNEY

SD	67% (33° 49')		68% (34° 13')		69% (34° 36')	
	HD	DE	HD	DE	HD	DE
35	29.08	19.48	28.94	19.68	28.81	19.88
40	33.23	22.26	33.08	22.49	32.92	22.72
45	37.38	25.05	37.21	25.30	37.04	25.56
50	41.54	27.83	41.35	28.12	41.15	28.40
55	45.69	30.61	45.48	30.93	45.27	31.24
60	49.85	33.40	49.62	33.74	49.38	34.08
65	54.00	36.18	53.75	36.55	53.50	36.92
70	58.15	38.96	57.88	39.36	57.62	39.75
75	62.31	41.75	62.02	42.17	61.73	42.59
80	66.46	44.53	66.15	44.98	65.85	45.43
85	70.62	47.31	70.29	47.80	69.96	48.27
90	74.77	50.10	74.42	50.61	74.08	51.11
95	78.92	52.88	78.56	53.42	78.19	53.95
100	83.08	55.66	82.69	56.23	82.31	56.79
105	87.23	58.44	86.83	59.04	86.42	59.63
110	91.38	61.23	90.96	61.85	90.54	62.47
115	95.54	64.01	95.10	64.67	94.65	65.31
120	99.69	66.79	99.23	67.48	98.77	68.15
125	103.85	69.58	103.37	70.29	102.89	70.99
130	108.00	72.36	107.50	73.10	107.00	73.83
135	112.15	75.14	111.64	75.91	111.12	76.67
140	116.31	77.93	115.77	78.72	115.23	79.51
145	120.46	80.71	119.90	81.53	119.35	82.35
150	124.62	83.49	124.04	84.35	123.46	85.19
155	128.77	86.28	128.17	87.16	127.58	88.03
160	132.92	89.06	132.31	89.97	131.69	90.87
165	137.08	91.84	136.44	92.78	135.81	93.71
170	141.23	94.62	140.58	95.59	139.92	96.55
175	145.38	97.41	144.71	98.40	144.04	99.39
180	149.54	100.19	148.85	101.22	148.15	102.23
185	153.69	102.97	152.98	104.03	152.27	105.07
190	157.85	105.76	157.12	106.84	156.39	107.91
195	162.00	108.54	161.25	109.65	160.50	110.75
200	166.15	111.32	165.39	112.46	164.62	113.59
205	170.31	114.11	169.52	115.27	168.73	116.42
210	174.46	116.89	173.65	118.09	172.85	119.26
215	178.62	119.67	177.79	120.90	176.96	122.10
220	182.77	122.46	181.92	123.71	181.08	124.94
225	186.92	125.24	186.06	126.52	185.19	127.78
230	191.08	128.02	190.19	129.33	189.31	130.62
235	195.23	130.80	194.33	132.14	193.42	133.46
240	199.38	133.59	198.46	134.95	197.54	136.30
245	203.54	136.37	202.60	137.77	201.65	139.14
250	207.69	139.15	206.73	140.58	205.77	141.98
255	211.85	141.94	210.87	143.39	209.89	144.82
260	216.00	144.72	215.00	146.20	214.00	147.66
265	220.15	147.50	219.14	149.01	218.12	150.50
270	224.31	150.29	223.27	151.82	222.23	153.34
275	228.46	153.07	227.40	154.64	226.35	156.18
280	232.62	155.85	231.54	157.45	230.46	159.02
285	236.77	158.64	235.67	160.26	234.58	161.86
290	240.92	161.42	239.81	163.07	238.69	164.70
295	245.08	164.20	243.94	165.88	242.81	167.54
300	249.23	166.98	248.08	168.69	246.92	170.38

SLOPE REDUCTION TABLE FOR THE PERCENT ABNEY

SD	70%(34° 68')		71%(35° 22')		72%(35° 45')	
	HD	DE	HD	DE	HD	DE
35	28.67	20.07	28.54	20.26	28.40	20.45
40	32.77	22.94	32.62	23.16	32.46	23.37
45	36.87	25.81	36.69	26.05	36.52	26.29
50	40.96	28.67	40.77	28.95	40.58	29.22
55	45.06	31.54	44.85	31.84	44.63	32.14
60	49.15	34.41	48.92	34.74	48.69	35.06
65	53.25	37.28	53.00	37.63	52.75	37.98
70	57.35	40.14	57.08	40.52	56.81	40.90
75	61.44	43.01	61.15	43.42	60.87	43.82
80	65.54	45.88	65.23	46.31	64.92	46.74
85	69.63	48.74	69.31	49.21	58.98	49.67
90	73.73	51.61	73.38	52.10	73.04	52.59
95	77.83	54.48	77.46	55.00	77.10	55.51
100	81.92	57.35	81.54	57.89	81.15	58.43
105	86.02	60.21	85.62	60.79	85.21	61.35
110	90.12	63.08	89.69	63.68	89.27	64.27
115	94.21	65.95	93.77	66.58	93.33	67.20
120	98.31	68.82	97.85	69.47	97.38	70.12
125	102.40	71.68	101.92	72.37	101.44	73.04
130	106.50	74.55	106.00	75.26	105.50	75.96
135	110.60	77.42	110.08	78.15	109.56	78.88
140	114.69	80.28	114.15	81.05	113.61	81.80
145	118.79	83.15	118.23	83.94	117.67	84.72
150	122.88	86.02	122.31	86.84	121.73	87.65
155	126.98	88.89	126.38	89.73	125.79	90.57
160	131.08	91.75	130.46	92.63	129.85	93.49
165	135.17	94.62	134.54	95.52	133.90	96.41
170	139.27	97.49	138.62	98.42	137.96	99.33
175	143.37	100.36	142.69	101.31	142.02	102.25
180	147.46	103.22	146.77	104.21	146.04	105.17
185	151.56	106.09	150.85	107.10	150.13	108.10
190	155.65	108.96	154.92	110.00	154.19	111.02
195	159.75	111.83	159.00	112.89	158.25	113.94
200	163.85	114.69	163.08	115.78	162.31	116.86
205	167.94	117.56	167.15	118.68	166.36	119.78
210	172.04	120.43	171.23	121.57	170.42	122.70
215	176.13	123.29	175.31	124.47	174.48	125.63
220	180.23	126.16	179.38	127.36	178.54	128.55
225	184.33	129.03	183.46	130.26	182.60	131.47
230	188.42	131.90	187.54	133.15	186.65	134.39
235	192.52	134.76	191.62	136.05	190.71	137.31
240	196.62	137.63	195.69	138.94	194.77	140.23
245	200.71	140.50	199.77	141.84	198.83	143.15
250	204.81	143.37	203.85	144.73	202.88	146.08
255	208.90	146.23	207.92	147.63	206.94	149.00
260	213.00	149.10	212.00	150.52	211.00	151.92
265	217.10	151.97	216.08	153.41	215.06	154.84
270	221.19	154.83	220.15	156.31	219.11	157.76
275	225.29	157.70	224.23	159.20	223.17	160.68
280	229.38	160.57	228.31	162.10	227.23	163.61
285	233.48	163.44	232.38	164.99	231.29	166.53
290	237.58	166.30	236.46	167.89	235.34	169.45
295	241.67	169.17	240.54	170.78	239.40	172.37
300	245.77	172.04	244.61	173.68	243.46	175.29

SLOPE REDUCTION TABLE FOR THE PERCENT ABNEY

	73%(36° 08′)		74%(36° 30′)		75%(36° 52′)	
SO	HD	OE	HD	OE	HD	OE
35	28.27	20.64	28.13	20.82	28.00	21.00
40	32.31	23.58	32.15	23.79	32.00	24.00
45	36.35	26.53	36.17	26.77	36.00	27.00
50	40.38	29.48	40.19	29.74	40.00	30.00
55	44.42	32.43	44.21	32.72	44.00	33.00
60	48.46	35.38	48.23	35.69	48.00	36.00
65	52.50	38.32	52.25	38.66	52.00	39.00
70	56.54	41.27	56.27	41.64	56.00	42.00
75	60.58	44.22	60.29	44.61	60.00	45.00
80	64.61	47.17	64.31	47.59	64.00	48.00
85	68.65	50.12	68.33	50.56	68.00	51.00
90	72.69	53.07	72.35	53.54	72.00	54.00
95	76.73	56.01	76.36	56.51	76.00	57.00
100	80.77	58.96	80.38	59.48	80.00	60.00
105	84.81	61.91	84.40	62.46	84.00	63.00
110	88.85	64.86	88.42	65.43	88.00	66.00
115	92.88	67.81	92.44	68.41	92.00	69.00
120	96.92	70.75	96.46	71.38	96.00	72.00
125	100.96	73.70	100.48	74.36	100.00	75.00
130	105.00	76.65	104.50	77.33	104.00	78.00
135	109.04	79.60	108.52	80.30	108.00	81.00
140	113.08	82.55	112.54	83.28	112.00	84.00
145	117.11	85.49	116.56	86.25	116.00	87.00
150	121.15	88.44	120.58	89.23	120.00	90.00
155	125.19	91.39	124.60	92.20	124.00	93.00
160	129.23	94.34	128.61	95.17	128.00	96.00
165	133.27	97.29	132.63	98.15	132.00	99.00
170	137.31	100.23	136.65	101.12	136.00	102.00
175	141.35	103.18	140.67	104.10	140.00	105.00
180	145.38	106.13	144.69	107.07	144.00	108.00
185	149.42	109.08	148.71	110.05	148.00	111.00
190	153.46	112.03	152.73	113.02	152.00	114.00
195	157.50	114.97	156.75	115.99	156.00	117.00
200	161.54	117.92	160.77	118.97	160.00	120.00
205	165.58	120.87	164.79	121.94	164.00	123.00
210	169.61	123.82	168.81	124.92	168.00	126.00
215	173.65	126.77	172.83	127.89	172.00	129.00
220	177.69	129.71	176.85	130.87	176.00	132.00
225	181.73	132.66	180.86	133.84	180.00	135.00
230	185.77	135.61	184.88	136.81	184.00	138.00
235	189.81	138.56	188.90	139.79	188.00	141.00
240	193.84	141.51	192.92	142.76	192.00	144.00
245	197.88	144.45	196.94	145.74	196.00	147.00
250	201.92	147.40	200.96	148.71	200.00	150.00
255	205.96	150.35	204.98	151.68	204.00	153.00
260	210.00	153.30	209.00	154.66	208.00	156.00
265	214.04	156.25	213.02	157.63	212.00	159.00
270	218.08	159.20	217.04	160.61	216.00	162.00
275	222.11	162.14	221.06	163.58	220.00	165.00
280	226.15	165.09	225.08	166.56	224.00	168.00
285	230.19	168.04	229.09	169.53	228.00	171.00
290	234.23	170.99	233.11	172.50	232.00	174.00
295	238.27	173.94	237.13	175.48	236.00	177.00
300	242.31	176.88	241.15	178.45	240.00	180.00

SLOPE REDUCTION TABLE FOR THE PERCENT ABNEY

SO	76%(37° 14') HD	76%(37° 14') DE	77%(37° 36') HD	77%(37° 36') DE	78%(37° 57') HD	78%(37° 57') DE
35	27.87	21.18	27.73	21.35	27.60	21.53
40	31.85	24.20	31.69	24.40	31.54	24.60
45	35.83	27.23	35.65	27.45	35.48	27.68
50	39.81	30.25	39.62	30.50	39.43	30.75
55	43.79	33.28	43.58	33.56	43.37	33.83
60	47.77	36.30	47.54	36.61	47.31	36.90
65	51.75	39.33	51.50	39.66	51.25	39.98
70	55.73	42.36	55.46	42.71	55.20	43.05
75	59.71	45.38	59.42	45.76	59.14	46.13
80	63.69	48.41	63.39	48.81	63.08	49.20
85	67.67	51.43	67.35	51.86	67.02	52.28
90	71.65	54.46	71.31	54.91	70.97	55.35
95	75.64	57.48	75.27	57.96	74.91	58.43
100	79.62	60.51	79.23	61.01	78.85	61.50
105	83.60	63.53	83.19	64.06	82.79	64.58
110	87.58	66.56	87.16	67.11	86.74	67.65
115	91.56	69.58	91.12	70.16	90.68	70.73
120	95.54	72.61	95.08	73.21	94.62	73.80
125	99.52	75.64	99.04	76.26	98.56	76.88
130	103.50	78.66	103.00	79.31	102.51	79.95
135	107.48	81.69	106.96	82.36	106.45	83.03
140	111.46	84.71	110.93	85.41	110.39	86.10
145	115.44	87.74	114.89	88.46	114.33	89.18
150	119.42	90.76	118.85	91.51	118.28	92.25
155	123.41	93.79	122.81	94.56	122.22	95.33
160	127.39	96.81	126.77	97.61	126.16	98.41
165	131.37	99.84	130.73	100.67	130.10	101.48
170	135.35	102.86	134.70	103.72	134.05	104.56
175	139.33	105.89	138.66	106.77	137.99	107.63
180	143.31	108.91	142.62	109.82	141.93	110.71
185	147.29	111.94	146.58	112.87	145.87	113.78
190	151.27	114.97	150.54	115.92	149.82	116.86
195	155.25	117.99	154.50	118.97	153.76	119.93
200	159.23	121.02	158.47	122.02	157.70	123.01
205	163.21	124.04	162.43	125.07	161.64	126.08
210	167.19	127.07	166.39	128.12	165.59	129.16
215	171.17	130.09	170.35	131.17	169.53	132.23
220	175.16	133.12	174.31	134.22	173.47	135.31
225	179.14	136.14	178.27	137.27	177.41	138.38
230	183.12	139.17	182.24	140.32	181.36	141.46
235	187.10	142.19	186.20	143.37	185.30	144.53
240	191.08	145.22	190.16	146.42	189.24	147.61
245	195.06	148.25	194.12	149.47	193.18	150.68
250	199.04	151.27	198.08	152.52	197.13	153.76
255	203.02	154.30	202.04	155.57	201.07	156.83
260	207.00	157.32	206.01	158.62	205.01	159.91
265	210.98	160.35	209.97	161.67	208.95	162.98
270	214.96	163.37	213.93	164.73	212.90	166.06
275	218.94	166.40	217.89	167.78	216.84	169.13
280	222.93	169.42	221.85	170.83	220.78	172.21
285	226.91	172.45	225.81	173.88	224.72	175.28
290	230.89	175.47	229.78	176.93	228.67	178.36
295	234.87	178.50	233.74	179.98	232.61	181.43
300	238.85	181.52	237.70	183.03	236.55	184.51

SLOPE REDUCTION TABLE FOR THE PERCENT ABNEY

	79%(38° 19′)		80%(38° 40′)		81%(39° 00′)	
SD	HD	DE	HD	DE	HD	DE
35	27.46	21.70	27.33	21.86	27.20	22.03
40	31.39	24.80	31.23	24.99	31.08	25.18
45	35.31	27.90	35.14	28.11	34.97	28.32
50	39.23	30.99	39.04	31.23	38.85	31.47
55	43.16	34.09	42.95	34.36	42.74	34.62
60	47.08	37.19	46.85	37.48	46.62	37.77
65	51.00	40.29	50.76	40.61	50.51	40.91
70	54.93	43.39	54.66	43.73	54.39	44.06
75	58.85	46.49	58.57	46.85	58.28	47.21
80	62.77	49.59	62.47	49.98	62.17	50.35
85	66.70	52.69	66.37	53.10	66.05	53.50
90	70.62	55.79	70.28	56.22	69.94	56.65
95	74.54	58.89	74.18	59.35	73.82	59.80
100	78.47	61.99	78.09	62.47	77.71	62.94
105	82.39	65.09	81.99	65.59	81.59	66.09
110	86.32	68.19	85.90	68.72	85.48	69.24
115	90.24	71.29	89.80	71.84	89.36	72.38
120	94.16	74.39	93.70	74.96	93.25	75.53
125	98.09	77.49	97.61	78.09	97.13	78.68
130	102.01	80.59	101.51	81.21	101.02	81.82
135	105.93	83.69	105.42	84.33	104.90	84.97
140	109.86	86.79	109.32	87.46	108.79	88.12
145	113.78	89.89	113.23	90.58	112.67	91.27
150	117.70	92.98	117.13	93.70	116.56	94.41
155	121.63	96.08	121.03	96.83	120.44	97.56
160	125.55	99.18	124.94	99.95	124.33	100.71
165	129.47	102.28	128.84	103.07	128.22	103.85
170	133.40	105.38	132.75	106.20	132.10	107.00
175	137.32	108.48	136.65	109.32	135.99	110.15
180	141.24	111.58	140.56	112.45	139.87	113.30
185	145.17	114.68	144.46	115.57	143.76	116.44
190	149.09	117.78	148.37	118.69	147.64	119.59
195	153.01	120.88	152.27	121.82	151.53	122.74
200	156.94	123.98	156.17	124.94	155.41	125.88
205	160.86	127.08	160.08	128.06	159.30	129.03
210	164.78	130.18	163.98	131.19	163.18	132.18
215	168.71	133.28	167.89	134.31	167.07	135.33
220	172.63	136.38	171.79	137.43	170.95	138.47
225	176.55	139.48	175.70	140.56	174.84	141.62
230	180.48	142.58	179.60	143.68	178.72	144.77
235	184.40	145.68	183.50	146.80	182.61	147.91
240	188.32	148.78	187.41	149.93	186.50	151.06
245	192.25	151.88	191.31	153.05	190.38	154.21
250	196.17	154.97	195.22	156.17	194.27	157.36
255	200.09	158.07	199.12	159.30	198.15	160.50
260	204.02	161.17	203.03	162.42	202.04	163.65
265	207.94	164.27	206.93	165.54	205.92	166.80
270	211.86	167.37	210.83	168.67	209.81	169.94
275	215.79	170.47	214.74	171.79	213.69	173.09
280	219.71	173.57	218.64	174.91	217.58	176.24
285	223.63	176.67	222.55	178.04	221.46	179.39
290	227.56	179.77	226.45	181.16	225.35	182.53
295	231.48	182.87	230.36	184.29	229.23	185.68
300	235.40	185.97	234.26	187.41	233.12	188.83

SLOPE REDUCTION TABLE FOR THE PERCENT ABNEY

SD	82% (39° 21')		83% (39° 42')		84% (40° 02')	
	HD	DE	HD	DE	HD	DE
35	27.06	22.19	26.93	22.35	26.80	22.51
40	30.93	25.36	30.78	25.55	30.63	25.73
45	34.80	28.53	34.63	28.74	34.46	28.94
50	38.66	31.70	38.47	31.93	38.29	32.16
55	42.53	34.87	42.32	35.13	42.11	35.38
60	46.40	38.04	46.17	38.32	45.94	38.59
65	50.26	41.22	50.02	41.51	49.77	41.81
70	54.13	44.39	53.86	44.71	53.60	45.02
75	58.00	47.56	57.71	47.90	57.43	48.24
80	61.86	50.73	61.56	51.09	61.26	51.46
85	65.73	53.90	65.41	54.29	65.08	54.67
90	69.59	57.07	69.25	57.48	68.91	57.89
95	73.46	60.24	73.10	60.67	72.74	61.10
100	77.33	63.41	76.95	63.87	76.57	64.32
105	81.19	66.58	80.80	67.06	80.40	67.54
110	85.06	69.75	84.64	70.25	84.23	70.75
115	88.93	72.92	88.49	73.45	88.06	73.97
120	92.79	76.09	92.34	76.64	91.88	77.18
125	96.66	79.26	96.19	79.83	95.71	80.40
130	100.52	82.43	100.03	83.03	99.54	83.61
135	104.39	85.60	103.88	86.22	103.37	86.83
140	108.26	88.77	107.73	89.41	107.20	90.05
145	112.12	91.94	111.57	92.61	111.03	93.26
150	115.99	95.11	115.42	95.80	114.86	96.48
155	119.86	98.28	119.27	98.99	118.68	99.69
160	123.72	101.45	123.12	102.19	122.51	102.91
165	127.59	104.62	126.96	105.38	126.34	106.13
170	131.46	107.79	130.81	108.57	130.17	109.34
175	135.32	110.96	134.66	111.77	134.00	112.56
180	139.19	114.13	138.51	114.96	137.83	115.77
185	143.05	117.30	142.35	118.15	141.66	118.99
190	146.92	120.48	146.20	121.35	145.48	122.21
195	150.79	123.65	150.05	124.54	149.31	125.42
200	154.65	126.82	153.90	127.73	153.14	128.64
205	158.52	129.99	157.74	130.93	156.97	131.85
210	162.39	133.16	161.59	134.12	160.80	135.07
215	166.25	136.33	165.44	137.31	164.63	138.29
220	170.12	139.50	169.29	140.51	168.46	141.50
225	173.99	142.67	173.13	143.70	172.28	144.72
230	177.85	145.84	176.98	146.89	176.11	147.93
235	181.72	149.01	180.83	150.09	179.94	151.15
240	185.58	152.18	184.68	153.28	183.77	154.37
245	189.45	155.35	188.52	156.47	187.60	157.58
250	193.32	158.52	192.37	159.67	191.43	160.80
255	197.18	161.69	196.22	162.86	195.25	164.01
260	201.05	164.86	200.07	166.05	199.08	167.23
265	204.92	168.03	203.91	169.25	202.91	170.45
270	208.78	171.20	207.76	172.44	206.74	173.66
275	212.65	174.37	211.61	175.63	210.57	176.88
280	216.51	177.54	215.45	178.83	214.40	180.09
285	220.38	180.71	219.30	182.02	218.23	183.31
290	224.25	183.88	223.15	185.21	222.05	186.53
295	228.11	187.05	227.00	188.41	225.88	189.74
300	231.98	190.22	230.84	191.60	229.71	192.96

SLOPE REDUCTION TABLE FOR THE PERCENT ABNEY

SD	85% (40° 22') HD	DE	86% (40° 42') HD	DE	87% (41° 01') HD	DE
35	26.67	22.67	26.54	22.82	26.41	22.97
40	30.48	25.91	30.33	26.08	30.18	26.25
45	34.29	29.14	34.12	29.34	33.95	29.54
50	38.10	32.38	37.91	32.60	37.72	32.82
55	41.91	35.62	41.70	35.86	41.49	36.10
60	45.72	38.86	45.49	39.12	45.27	39.38
65	49.53	42.10	49.28	42.38	49.04	42.66
70	53.34	45.34	53.07	45.64	52.81	45.95
75	57.15	48.57	56.86	48.90	56.58	49.23
80	60.96	51.81	60.65	52.16	60.36	52.51
85	64.76	55.05	64.45	55.42	64.13	55.79
90	68.57	58.29	68.24	58.68	67.90	59.07
95	72.38	61.53	72.03	61.94	71.67	62.35
100	76.19	64.76	75.82	65.20	75.44	65.64
105	80.00	68.00	79.61	68.46	79.22	68.92
110	83.81	71.24	83.40	71.72	82.99	72.20
115	87.62	74.48	87.19	74.98	86.76	75.48
120	91.43	77.72	90.98	78.24	90.53	78.76
125	95.24	80.96	94.77	81.50	94.31	82.05
130	99.05	84.19	98.56	84.77	98.08	85.33
135	102.86	87.43	102.36	88.03	101.85	88.61
140	106.67	90.67	106.15	91.29	105.62	91.89
145	110.48	93.91	109.94	94.55	109.39	95.17
150	114.29	97.15	113.73	97.81	113.17	98.45
155	118.10	100.39	117.52	101.07	116.94	101.74
160	121.91	103.62	121.31	104.33	120.71	105.02
165	125.72	106.86	125.10	107.59	124.48	108.30
170	129.53	110.10	128.89	110.85	128.26	111.58
175	133.34	113.34	132.68	114.11	132.03	114.86
180	137.15	116.58	136.47	117.37	135.80	118.15
185	140.96	119.81	140.26	120.63	139.57	121.43
190	144.77	123.05	144.06	123.89	143.34	124.71
195	148.58	126.29	147.85	127.15	147.12	127.99
200	152.39	129.53	151.64	130.41	150.89	131.27
205	156.20	132.77	155.43	133.67	154.66	134.55
210	160.01	136.01	159.22	136.93	158.43	137.84
215	163.82	139.24	163.01	140.19	162.21	141.12
220	167.63	142.48	166.80	143.45	165.98	144.40
225	171.44	145.72	170.59	146.71	169.75	147.68
230	175.25	148.96	174.38	149.97	173.52	150.96
235	179.06	152.20	178.17	153.23	177.29	154.25
240	182.87	155.44	181.96	156.49	181.07	157.53
245	186.68	158.67	185.76	159.75	184.84	160.81
250	190.48	161.91	189.55	163.01	188.61	164.09
255	194.29	165.15	193.34	166.27	192.38	167.37
260	198.10	168.39	197.13	169.53	196.16	170.66
265	201.91	171.63	200.92	172.79	199.93	173.94
270	205.72	174.87	204.71	176.05	203.70	177.22
275	209.53	178.10	208.50	179.31	207.47	180.50
280	213.34	181.34	212.29	182.57	211.24	183.78
285	217.15	184.58	216.08	185.83	215.02	187.06
290	220.96	187.82	219.87	189.09	218.79	190.35
295	224.77	191.06	223.66	192.35	222.56	193.63
300	228.58	194.29	227.46	195.61	226.33	196.91

SLOPE REDUCTION TABLE FOR THE PERCENT ABNEY

	88%(41° 21')		89%(41° 40')		90%(41° 59')	
SD	HD	DE	HD	DE	HD	DE
35	26.27	23.12	26.14	23.27	26.02	23.41
40	30.03	26.43	29.88	26.59	29.73	26.76
45	33.78	29.73	33.61	29.92	33.45	30.10
50	37.54	33.03	37.35	33.24	37.16	33.45
55	41.29	36.33	41.08	36.57	40.88	36.79
60	45.04	39.64	44.82	39.89	44.60	40.14
65	48.80	42.94	48.55	43.21	48.31	43.48
70	52.55	46.24	52.29	46.54	52.03	46.83
75	56.30	49.55	56.02	49.86	55.75	50.17
80	60.06	52.85	59.76	53.19	59.46	53.52
85	63.81	56.15	63.49	56.51	63.18	56.86
90	67.56	59.46	67.23	59.83	66.90	60.21
95	71.32	62.76	70.96	63.16	70.61	63.55
100	75.07	66.06	74.70	66.48	74.33	66.90
105	78.82	69.37	78.43	69.81	78.05	70.24
110	82.58	72.67	82.17	73.13	81.76	73.59
115	86.33	75.97	85.90	76.46	85.48	76.93
120	90.09	79.28	89.64	79.78	89.20	80.28
125	93.84	82.58	93.37	83.10	92.91	83.62
130	97.59	85.88	97.11	86.43	96.63	86.97
135	101.35	89.18	100.84	89.75	100.34	90.31
140	105.10	92.49	104.58	93.08	104.06	93.66
145	108.85	95.79	108.31	96.40	107.78	97.00
150	112.61	99.09	112.05	99.72	111.49	100.34
155	116.36	102.40	115.78	103.05	115.21	103.69
160	120.11	105.70	119.52	106.37	118.93	107.03
165	123.87	109.00	123.25	109.70	122.64	110.38
170	127.62	112.31	126.99	113.02	126.36	113.72
175	131.37	115.61	130.72	116.34	130.08	117.07
180	135.13	118.91	134.46	119.67	133.79	120.41
185	138.88	122.22	138.19	122.99	137.51	123.76
190	142.64	125.52	141.93	126.32	141.23	127.10
195	146.39	128.82	145.66	129.64	144.94	130.45
200	150.14	132.13	149.40	132.97	148.66	133.79
205	153.90	135.43	153.13	136.29	152.38	137.14
210	157.65	138.73	156.87	139.61	156.09	140.48
215	161.40	142.03	160.60	142.94	159.81	143.83
220	165.16	145.34	164.34	146.26	163.52	147.17
225	168.91	148.64	168.07	149.59	167.24	150.52
230	172.66	151.94	171.81	152.91	170.96	153.86
235	176.42	155.25	175.54	156.23	174.67	157.21
240	180.17	158.55	179.28	159.56	178.39	160.55
245	183.92	161.85	183.01	162.88	182.11	163.90
250	187.68	165.16	186.75	166.21	185.82	167.24
255	191.43	168.46	190.48	169.53	189.54	170.59
260	195.19	171.76	194.22	172.86	193.26	173.93
265	198.94	175.07	197.95	176.18	196.97	177.28
270	202.69	178.37	201.69	179.50	200.69	180.62
275	206.45	181.67	205.42	182.83	204.41	183.97
280	210.20	184.98	209.16	186.15	208.12	187.31
285	213.95	188.28	212.89	189.48	211.84	190.65
290	217.71	191.58	216.63	192.80	215.56	194.00
295	221.46	194.89	220.36	196.12	219.27	197.34
300	225.21	198.19	224.10	199.45	222.99	200.69

SLOPE REDUCTION TABLE FOR THE PERCENT ABNEY

SD	91%(42° 18')		92%(42° 37')		93%(42° 55')	
	HD	DE	HD	DE	HD	DE
35	25.89	23.56	25.76	23.70	25.63	23.84
40	29.58	26.92	29.44	27.08	29.29	27.24
45	33.28	30.29	33.12	30.47	32.95	30.65
50	36.98	33.65	36.80	33.85	36.61	34.05
55	40.68	37.02	40.48	37.24	40.27	37.46
60	44.38	40.38	44.16	40.62	43.94	40.86
65	48.07	43.75	47.84	44.01	47.60	44.27
70	51.77	47.11	51.52	47.39	51.26	47.67
75	55.47	50.48	55.19	50.78	54.92	51.08
80	59.17	53.84	58.87	54.16	58.58	54.48
85	62.87	57.21	62.55	57.55	62.24	57.89
90	66.56	60.57	66.23	60.94	65.90	61.29
95	70.26	63.94	69.91	64.32	69.57	64.70
100	73.96	67.30	73.59	67.71	73.23	68.10
105	77.66	70.67	77.27	71.09	76.89	71.51
110	81.36	74.03	80.95	74.48	80.55	74.91
115	85.05	77.40	84.63	77.86	84.21	78.32
120	88.75	80.76	88.31	81.25	87.87	81.72
125	92.45	84.13	91.99	84.63	91.53	85.13
130	96.15	87.50	95.67	88.02	95.20	88.53
135	99.85	90.86	99.35	91.40	98.86	91.94
140	103.54	94.23	103.03	94.79	102.52	95.34
145	107.24	97.59	106.71	98.17	106.18	98.75
150	110.94	100.96	110.39	101.56	109.84	102.15
155	114.64	104.32	114.07	104.94	113.50	105.56
160	118.34	107.69	117.75	108.33	117.16	108.96
165	122.03	111.05	121.43	111.71	120.82	112.37
170	125.73	114.42	125.11	115.10	124.49	115.77
175	129.43	117.78	128.79	118.48	128.15	119.18
180	133.13	121.15	132.47	121.87	131.81	122.58
185	136.83	124.51	136.15	125.26	135.47	125.99
190	140.53	127.88	139.83	128.64	139.13	129.39
195	144.22	131.24	143.51	132.03	142.79	132.80
200	147.92	134.61	147.19	135.41	146.45	136.20
205	151.62	137.97	150.87	138.80	150.12	139.61
210	155.32	141.34	154.55	142.18	153.78	143.01
215	159.02	144.70	158.23	145.57	157.44	146.42
220	162.71	148.07	161.90	148.95	161.10	149.82
225	166.41	151.43	165.58	152.34	164.76	153.23
230	170.11	154.80	169.26	155.72	168.42	156.63
235	173.81	158.16	172.94	159.11	172.08	160.04
240	177.51	161.53	176.62	162.49	175.75	163.44
245	181.20	164.90	180.30	165.88	179.41	166.85
250	184.90	168.26	183.98	169.26	183.07	170.25
255	188.60	171.63	187.66	172.65	186.73	173.66
260	192.30	174.99	191.34	176.03	190.39	177.06
265	196.00	178.36	195.02	179.42	194.05	180.47
270	199.69	181.72	198.70	182.81	197.71	183.87
275	203.39	185.09	202.38	186.19	201.37	187.28
280	207.09	188.45	206.06	189.58	205.04	190.68
285	210.79	191.82	209.74	192.96	208.70	194.09
290	214.49	195.18	213.42	196.35	212.36	197.49
295	218.18	198.55	217.10	199.73	216.02	200.90
300	221.88	201.91	220.78	203.12	219.68	204.30

SLOPE REDUCTION TABLE FOR THE PERCENT ABNEY

SD	94%(43°14') HD	DE	95%(43°32') HD	DE	96%(43°50') HD	DE
35	25.50	23.97	25.37	24.11	25.25	24.24
40	29.15	27.40	29.00	27.55	28.86	27.70
45	32.79	30.82	32.62	30.99	32.46	31.16
50	36.43	34.25	36.25	34.44	36.07	34.63
55	40.07	37.67	39.87	37.88	39.68	38.09
60	43.72	41.09	43.50	41.32	43.28	41.55
65	47.36	44.52	47.12	44.77	46.89	45.01
70	51.00	47.94	50.75	48.21	50.50	48.48
75	54.65	51.37	54.37	51.66	54.10	51.94
80	58.29	54.79	58.00	55.10	57.71	55.40
85	61.93	58.22	61.62	58.54	61.32	58.87
90	65.58	61.64	65.25	61.99	64.92	62.33
95	69.22	65.07	68.87	65.43	68.53	65.79
100	72.86	68.49	72.50	68.87	72.14	69.25
105	76.51	71.92	76.12	72.32	75.75	72.72
110	80.15	75.34	79.75	75.76	79.35	76.18
115	83.79	78.76	83.37	79.21	82.96	79.64
120	87.44	82.19	87.00	82.65	86.57	83.10
125	91.08	85.61	90.62	86.09	90.17	86.57
130	94.72	89.04	94.25	89.54	93.78	90.03
135	98.36	92.46	97.87	92.98	97.39	93.49
140	102.01	95.89	101.50	96.42	100.99	96.95
145	105.65	99.31	105.12	99.87	104.60	100.42
150	109.29	102.74	108.75	103.31	108.21	103.88
155	112.94	106.16	112.37	106.76	111.82	107.34
160	116.58	109.59	116.00	110.20	115.42	110.81
165	120.22	113.01	119.62	113.64	119.03	114.27
170	123.87	116.43	123.25	117.09	122.64	117.73
175	127.51	119.86	126.87	120.53	126.24	121.19
180	131.15	123.28	130.50	123.97	129.85	124.66
185	134.80	126.71	134.12	127.42	133.46	128.12
190	138.44	130.13	137.75	130.86	137.06	131.58
195	142.08	133.56	141.37	134.31	140.67	135.04
200	145.73	136.98	145.00	137.75	144.28	138.51
205	149.37	140.41	148.62	141.19	147.88	141.97
210	153.01	143.83	152.25	144.64	151.49	145.43
215	156.65	147.26	155.87	148.08	155.10	148.89
220	160.30	150.68	159.50	151.52	158.71	152.36
225	163.94	154.10	163.12	154.97	162.31	155.82
230	167.58	157.53	166.75	158.41	165.92	159.28
235	171.23	160.95	170.37	161.86	169.53	162.74
240	174.87	164.38	174.00	165.30	173.13	166.21
245	178.51	167.80	177.62	168.74	176.74	169.67
250	182.16	171.23	181.25	172.19	180.35	173.13
255	185.80	174.65	184.87	175.63	183.95	176.60
260	189.44	178.08	188.50	179.07	187.56	180.06
265	193.09	181.50	192.12	182.52	191.17	183.52
270	196.73	184.93	195.75	185.96	194.77	186.98
275	200.37	188.35	199.37	189.41	198.38	190.45
280	204.02	191.77	203.00	192.85	201.99	193.91
285	207.66	195.20	206.62	196.29	205.60	197.37
290	211.30	198.62	210.25	199.74	209.20	200.83
295	214.95	202.05	213.87	203.18	212.81	204.30
300	218.59	205.47	217.50	206.62	216.42	207.76

SLOPE REDUCTION TABLE FOR THE PERCENT ABNEY

SD	97%(44°08') HD	DE	98%(44°25') HD	DE	99%(44°43') HD	DE
35	25.12	24.37	25.00	24.50	24.87	24.62
40	28.71	27.85	28.57	28.00	28.43	28.14
45	32.30	31.33	32.14	31.50	31.98	31.66
50	35.89	34.81	35.71	35.00	35.53	35.18
55	39.48	38.29	39.28	38.50	39.09	38.69
60	43.07	41.78	42.85	42.00	42.64	42.21
65	46.66	45.26	46.42	45.50	46.19	45.73
70	50.25	48.74	49.99	49.00	49.75	49.25
75	53.83	52.22	53.57	52.49	53.30	52.77
80	57.42	55.70	57.14	55.99	56.85	56.28
85	61.01	59.18	60.71	59.49	60.41	59.80
90	64.60	62.66	64.28	62.99	63.96	63.32
95	68.19	66.14	67.85	66.49	67.51	66.84
100	71.78	69.63	71.42	69.99	71.07	70.35
105	75.37	73.11	74.99	73.49	74.62	73.87
110	78.96	76.59	78.56	76.99	78.17	77.39
115	82.55	80.07	82.13	80.49	81.72	80.91
120	86.14	83.55	85.71	83.99	85.29	84.43
125	89.72	87.03	89.28	87.49	88.83	87.94
130	93.31	90.51	92.85	90.99	92.38	91.46
135	96.90	93.99	96.42	94.49	95.94	94.98
140	100.49	97.48	99.99	97.99	99.49	98.50
145	104.08	100.96	103.56	101.49	103.04	102.01
150	107.67	104.44	107.13	104.99	106.60	105.53
155	111.26	107.92	110.70	108.49	110.15	109.05
160	114.85	111.40	114.27	111.99	113.70	112.57
165	118.44	114.88	117.85	115.49	117.26	116.08
170	122.02	118.36	121.42	118.99	120.81	119.60
175	125.61	121.85	124.99	122.49	124.36	123.12
180	129.20	125.33	128.56	125.99	127.92	126.64
185	132.79	128.81	132.13	129.49	131.47	130.16
190	136.38	132.29	135.70	132.99	135.02	133.67
195	139.97	135.77	139.27	136.49	138.58	137.19
200	143.56	139.25	142.84	139.99	142.13	140.71
205	147.15	142.73	146.41	143.49	145.68	144.23
210	150.74	146.21	149.98	146.99	149.24	147.74
215	154.33	149.70	153.56	150.48	152.79	151.26
220	157.91	153.18	157.13	153.98	156.34	154.78
225	161.50	156.66	160.70	157.48	159.90	158.30
230	165.09	160.14	164.27	160.98	163.45	161.82
235	168.68	163.62	167.84	164.48	167.00	165.33
240	172.27	167.10	171.41	167.98	170.56	168.85
245	175.86	170.58	174.98	171.48	174.11	172.37
250	179.45	174.06	178.55	174.98	177.66	175.89
255	183.04	177.55	182.12	178.48	181.22	179.40
260	186.63	181.03	185.70	181.98	184.77	182.92
265	190.21	184.51	189.27	185.48	188.32	186.44
270	193.80	187.99	192.84	188.98	191.88	189.96
275	197.39	191.47	196.41	192.48	195.43	193.47
280	200.98	194.95	199.98	195.99	198.98	196.99
285	204.57	198.43	203.55	199.48	202.54	200.51
290	208.16	201.91	207.12	202.98	206.09	204.03
295	211.75	205.40	210.69	206.48	209.64	207.55
300	215.34	208.88	214.26	209.98	213.20	211.06

Table 8. Natural trigonometric functions.

Values of the trigonometric functions of angles for each minute from $0^\circ-360^\circ$. For degrees indicated at the top of the page use the column headings at the top. For degrees indicated at the bottom use the column indications at the bottom.

With degrees at the left of each block (top or bottom), use the minute column at the left and with degrees at the right of each block use the minute column at the right.

These tables may also be used to compute latitudes and departures of courses when the distance and azimuth are known. Use the sine and cosine values of the angular value of the azimuth as directed above and then apply the correct sign depending in which quadrant the azimuth lies. For example the angular values of 32° and 57° are shown in the tables as follows:

32° (212°) (327°) 147°

122° (302°) (237°) 57°

Using the proper trigonometric value to compute the latitudes and departures, the following signs would be applied to the computed values.

32° (lat.+, dep.+), (212°)(lat.−, dep.−)
147° (lat.−, dep.+), (327°)(lat.+, dep.−)
57° (lat.+, dep.+), (237°)(lat.−, dep.−)
122° (lat.−, dep.+), (302°)(lat.+, dep.−)

0° (180°)   **(359°) 179°**

′	Sin	Tan	Cot	Cos	′
0	.00000	.00000	———	1.0000	60
1	.00029	.00029	3437.7	1.0000	59
2	.00058	.00058	1718.9	1.0000	58
3	.00087	.00087	1145.9	1.0000	57
4	.00116	.00116	859.44	1.0000	56
5	.00145	.00145	687.55	1.0000	55
6	.00175	.00175	572.96	1.0000	54
7	.00204	.00204	491.11	1.0000	53
8	.00233	.00233	429.72	1.0000	52
9	.00262	.00262	381.97	1.0000	51
10	.00291	.00291	343.77	1.0000	50
11	.00320	.00320	312.52	.99999	49
12	.00349	.00349	286.48	.99999	48
13	.00378	.00378	264.44	.99999	47
14	.00407	.00407	245.55	.99999	46
15	.00436	.00436	229.18	.99999	45
16	.00465	.00465	214.86	.99999	44
17	.00495	.00495	202.22	.99999	43
18	.00524	.00524	190.98	.99999	42
19	.00553	.00553	180.93	.99998	41
20	.00582	.00582	171.89	.99998	40
21	.00611	.00611	163.70	.99998	39
22	.00640	.00640	156.26	.99998	38
23	.00669	.00669	149.47	.99998	37
24	.00698	.00698	143.24	.99998	36
25	.00727	.00727	137.51	.99997	35
26	.00756	.00756	132.22	.99997	34
27	.00785	.00785	127.32	.99997	33
28	.00814	.00815	122.77	.99997	32
29	.00844	.00844	118.54	.99996	31
30	.00873	.00873	114.59	.99996	30
31	.00902	.00902	110.89	.99996	29
32	.00931	.00931	107.43	.99996	28
33	.00960	.00960	104.17	.99995	27
34	.00989	.00989	101.11	.99995	26
35	.01018	.01018	98.218	.99995	25
36	.01047	.01047	95.489	.99995	24
37	.01076	.01076	92.908	.99994	23
38	.01105	.01105	90.463	.99994	22
39	.01134	.01135	88.144	.99994	21
40	.01164	.01164	85.940	.99993	20
41	.01193	.01193	83.844	.99993	19
42	.01222	.01222	81.847	.99993	18
43	.01251	.01251	79.943	.99992	17
44	.01280	.01280	78.126	.99992	16
45	.01309	.01309	76.390	.99991	15
46	.01338	.01338	74.729	.99991	14
47	.01367	.01367	73.139	.99991	13
48	.01396	.01396	71.615	.99990	12
49	.01425	.01425	70.153	.99990	11
50	.01454	.01455	68.750	.99989	10
51	.01483	.01484	67.402	.99989	9
52	.01513	.01513	66.105	.99989	8
53	.01542	.01542	64.858	.99988	7
54	.01571	.01571	63.657	.99988	6
55	.01600	.01600	62.499	.99987	5
56	.01629	.01629	61.383	.99987	4
57	.01658	.01658	60.306	.99986	3
58	.01687	.01687	59.266	.99986	2
59	.01716	.01716	58.261	.99985	1
60	.01745	.01746	57.290	.99985	0
′	Cos	Cot	Tan	Sin	′

90° (270°)   **(269°) 89°**

1° (181°)   **(358°) 178°**

′	Sin	Tan	Cot	Cos	′
0	.01745	.01746	57.290	.99985	60
1	.01774	.01775	56.351	.99984	59
2	.01803	.01804	55.442	.99984	58
3	.01832	.01833	54.561	.99983	57
4	.01862	.01862	53.709	.99983	56
5	.01891	.01891	52.882	.99982	55
6	.01920	.01920	52.081	.99982	54
7	.01949	.01949	51.303	.99981	53
8	.01978	.01978	50.549	.99980	52
9	.02007	.02007	49.816	.99980	51
10	.02036	.02036	49.104	.99979	50
11	.02065	.02066	48.412	.99979	49
12	.02094	.02095	47.740	.99978	48
13	.02123	.02124	47.085	.99977	47
14	.02152	.02153	46.449	.99977	46
15	.02181	.02182	45.829	.99976	45
16	.02211	.02211	45.226	.99976	44
17	.02240	.02240	44.639	.99975	43
18	.02269	.02269	44.066	.99974	42
19	.02298	.02298	43.508	.99974	41
20	.02327	.02328	42.964	.99973	40
21	.02356	.02357	42.433	.99972	39
22	.02385	.02386	41.916	.99972	38
23	.02414	.02415	41.411	.99971	37
24	.02443	.02444	40.917	.99970	36
25	.02472	.02473	40.436	.99969	35
26	.02501	.02502	39.965	.99969	34
27	.02530	.02531	39.506	.99968	33
28	.02560	.02560	39.057	.99967	32
29	.02589	.02589	38.618	.99966	31
30	.02618	.02619	38.188	.99966	30
31	.02647	.02648	37.769	.99965	29
32	.02676	.02677	37.358	.99964	28
33	.02705	.02706	36.956	.99963	27
34	.02734	.02735	36.563	.99963	26
35	.02763	.02764	36.178	.99962	25
36	.02792	.02793	35.801	.99961	24
37	.02821	.02822	35.431	.99960	23
38	.02850	.02851	35.070	.99959	22
39	.02879	.02881	34.715	.99959	21
40	.02908	.02910	34.368	.99958	20
41	.02938	.02939	34.027	.99957	19
42	.02967	.02968	33.694	.99956	18
43	.02996	.02997	33.366	.99955	17
44	.03025	.03026	33.045	.99954	16
45	.03054	.03055	32.730	.99953	15
46	.03083	.03084	32.421	.99952	14
47	.03112	.03114	32.118	.99952	13
48	.03141	.03143	31.821	.99951	12
49	.03170	.03172	31.528	.99950	11
50	.03199	.03201	31.242	.99949	10
51	.03228	.03230	30.960	.99948	9
52	.03257	.03259	30.683	.99947	8
53	.03286	.03288	30.412	.99946	7
54	.03316	.03317	30.145	.99945	6
55	.03345	.03346	29.882	.99944	5
56	.03374	.03376	29.624	.99943	4
57	.03403	.03405	29.371	.99942	3
58	.03432	.03434	29.122	.99941	2
59	.03461	.03463	28.877	.99940	1
60	.03490	.03492	28.636	.99939	0
′	Cos	Cot	Tan	Sin	′

91° (271°)   **(268°) 88°**

2° (182°) (357°) 177° 3° (183°) (356°) 176°

′	Sin	Tan	Cot	Cos	′	′	Sin	Tan	Cot	Cos	′
0	.03490	.03492	28.636	.99939	60	0	.05234	.05241	19.081	.99863	60
1	.03519	.03521	28.399	.99938	59	1	.05263	.05270	18.976	.99861	59
2	.03548	.03550	28.166	.99937	58	2	.05292	.05299	18.871	.99860	58
3	.03577	.03579	27.937	.99936	57	3	.05321	.05328	18.768	.99858	57
4	.03606	.03609	27.712	.99935	56	4	.05350	.05357	18.666	.99857	56
5	.03635	.03638	27.490	.99934	55	5	.05379	.05387	18.564	.99855	55
6	.03664	.03667	27.271	.99933	54	6	.05408	.05416	18.464	.99854	54
7	.03693	.03696	27.057	.99932	53	7	.05437	.05445	18.366	.99852	53
8	.03723	.03725	26.845	.99931	52	8	.05466	.05474	18.268	.99851	52
9	.03752	.03754	26.637	.99930	51	9	.05495	.05503	18.171	.99849	51
10	.03781	.03783	26.432	.99929	50	10	.05524	.05533	18.075	.99847	50
11	.03810	.03812	26.230	.99927	49	11	.05553	.05562	17.980	.99846	49
12	.03839	.03842	26.031	.99926	48	12	.05582	.05591	17.886	.99844	48
13	.03868	.03871	25.835	.99925	47	13	.05611	.05620	17.793	.99842	47
14	.03897	.03900	25.642	.99924	46	14	.05640	.05649	17.702	.99841	46
15	.03926	.03929	25.452	.99923	45	15	.05669	.05678	17.611	.99839	45
16	.03955	.03958	25.264	.99922	44	16	.05698	.05708	17.521	.99838	44
17	.03984	.03987	25.080	.99921	43	17	.05727	.05737	17.431	.99836	43
18	.04013	.04016	24.898	.99919	42	18	.05756	.05766	17.343	.99834	42
19	.04042	.04046	24.719	.99918	41	19	.05785	.05795	17.256	.99833	41
20	.04071	.04075	24.542	.99917	40	20	.05814	.05824	17.169	.99831	40
21	.04100	.04104	24.368	.99916	39	21	.05844	.05854	17.084	.99829	39
22	.04129	.04133	24.196	.99915	38	22	.05873	.05883	16.999	.99827	38
23	.04159	.04162	24.026	.99913	37	23	.05902	.05912	16.915	.99826	37
24	.04188	.04191	23.859	.99912	36	24	.05931	.05941	16.832	.99824	36
25	.04217	.04220	23.695	.99911	35	25	.05960	.05970	16.750	.99822	35
26	.04246	.04250	23.532	.99910	34	26	.05989	.05999	16.668	.99821	34
27	.04275	.04279	23.372	.99909	33	27	.06018	.06029	16.587	.99819	33
28	.04304	.04308	23.214	.99907	32	28	.06047	.06058	16.507	.99817	32
29	.04333	.04337	23.058	.99906	31	29	.06076	.06087	16.428	.99815	31
30	.04362	.04366	22.904	.99905	30	30	.06105	.06116	16.350	.99813	30
31	.04391	.04395	22.752	.99904	29	31	.06134	.06145	16.272	.99812	29
32	.04420	.04424	22.602	.99902	28	32	.06163	.06175	16.195	.99810	28
33	.04449	.04454	22.454	.99901	27	33	.06192	.06204	16.119	.99808	27
34	.04478	.04483	22.308	.99900	26	34	.06221	.06233	16.043	.99806	26
35	.04507	.04512	22.164	.99898	25	35	.06250	.06262	15.969	.99804	25
36	.04536	.04541	22.022	.99897	24	36	.06279	.06291	15.895	.99803	24
37	.04565	.04570	21.881	.99896	23	37	.06308	.06321	15.821	.99801	23
38	.04594	.04599	21.743	.99894	22	38	.06337	.06350	15.748	.99799	22
39	.04623	.04628	21.606	.99893	21	39	.06366	.06379	15.676	.99797	21
40	.04653	.04658	21.470	.99892	20	40	.06395	.06408	15.605	.99795	20
41	.04682	.04687	21.337	.99890	19	41	.06424	.06438	15.534	.99793	19
42	.04711	.04716	21.205	.99889	18	42	.06453	.06467	15.464	.99792	18
43	.04740	.04745	21.075	.99888	17	43	.06482	.06496	15.394	.99790	17
44	.04769	.04774	20.946	.99886	16	44	.06511	.06525	15.325	.99788	16
45	.04798	.04803	20.819	.99885	15	45	.06540	.06554	15.257	.99786	15
46	.04827	.04833	20.693	.99883	14	46	.06569	.06584	15.189	.99784	14
47	.04856	.04862	20.569	.99882	13	47	.06598	.06613	15.122	.99782	13
48	.04885	.04891	20.446	.99881	12	48	.06627	.06642	15.056	.99780	12
49	.04914	.04920	20.325	.99879	11	49	.06656	.06671	14.990	.99778	11
50	.04943	.04949	20.206	.99878	10	50	.06685	.06700	14.924	.99776	10
51	.04972	.04978	20.087	.99876	9	51	.06714	.06730	14.860	.99774	9
52	.05001	.05007	19.970	.99875	8	52	.06743	.06759	14.795	.99772	8
53	.05030	.05037	19.855	.99873	7	53	.06773	.06788	14.732	.99770	7
54	.05059	.05066	19.740	.99872	6	54	.06802	.06817	14.669	.99768	6
55	.05088	.05095	19.627	.99870	5	55	.06831	.06847	14.606	.99766	5
56	.05117	.05124	19.516	.99869	4	56	.06860	.06876	14.544	.99764	4
57	.05146	.05153	19.405	.99867	3	57	.06889	.06905	14.482	.99762	3
58	.05175	.05182	19.296	.99866	2	58	.06918	.06934	14.421	.99760	2
59	.05205	.05212	19.188	.99864	1	59	.06947	.06963	14.361	.99758	1
60	.05234	.05241	19.081	.99863	0	60	.06976	.06993	14.301	.99756	0
′	Cos	Cot	Tan	Sin	′	′	Cos	Cot	Tan	Sin	′

4° (184°)   **(355°) 176°**

′	Sin	Tan	Cot	Cos	′
0	.06976	.06993	14.301	.99756	60
1	.07005	.07022	14.241	.99754	59
2	.07034	.07061	14.182	.99752	58
3	.07063	.07080	14.124	.99750	57
4	.07092	.07110	14.065	.99748	56
5	.07121	.07139	14.008	.99746	55
6	.07150	.07168	13.951	.99744	54
7	.07179	.07197	13.894	.99742	53
8	.07208	.07227	13.838	.99740	52
9	.07237	.07256	13.782	.99738	51
10	.07266	.07285	13.727	.99736	50
11	.07295	.07314	13.672	.99734	49
12	.07324	.07344	13.617	.99731	48
13	.07353	.07373	13.563	.99729	47
14	.07382	.07402	13.510	.99727	46
15	.07411	.07431	13.457	.99725	45
16	.07440	.07461	13.404	.99723	44
17	.07469	.07490	13.352	.99721	43
18	.07498	.07519	13.300	.99719	42
19	.07527	.07548	13.248	.99716	41
20	.07556	.07578	13.197	.99714	40
21	.07585	.07607	13.146	.99712	39
22	.07614	.07636	13.096	.99710	38
23	.07643	.07665	13.046	.99708	37
24	.07672	.07695	12.996	.99705	36
25	.07701	.07724	12.947	.99703	35
26	.07730	.07753	12.898	.99701	34
27	.07759	.07782	12.850	.99699	33
28	.07788	.07812	12.801	.99696	32
29	.07817	.07841	12.754	.99694	31
30	.07846	.07870	12.706	.99692	30
31	.07875	.07899	12.659	.99689	29
32	.07904	.07929	12.612	.99687	28
33	.07933	.07958	12.566	.99685	27
34	.07962	.07987	12.520	.99683	26
35	.07991	.08017	12.474	.99680	25
36	.08020	.08046	12.429	.99678	24
37	.08049	.08075	12.384	.99676	23
38	.08078	.08104	12.339	.99673	22
39	.08107	.08134	12.295	.99671	21
40	.08136	.08163	12.251	.99668	20
41	.08165	.08192	12.207	.99666	19
42	.08194	.08221	12.163	.99664	18
43	.08223	.08251	12.120	.99661	17
44	.08252	.08280	12.077	.99659	16
45	.08281	.08309	12.035	.99657	15
46	.08310	.08339	11.992	.99654	14
47	.08339	.08368	11.950	.99652	13
48	.08368	.08397	11.909	.99649	12
49	.08397	.08427	11.867	.99647	11
50	.08426	.08456	11.826	.99644	10
51	.08455	.08485	11.785	.99642	9
52	.08484	.08514	11.745	.99639	8
53	.08513	.08544	11.705	.99637	7
54	.08542	.08573	11.664	.99635	6
55	.08571	.08602	11.625	.99632	5
56	.08600	.08632	11.585	.99630	4
57	.08629	.08661	11.546	.99627	3
58	.08658	.08690	11.507	.99625	2
59	.08687	.08720	11.468	.99622	1
60	.08716	.08749	11.430	.99619	0
′	Cos	Cot	Tan	Sin	′

94° (274°)   **(265°) 85°**

5° (185°)   **(354°) 174°**

′	Sin	Tan	Cot	Cos	′
0	.08716	.08749	11.430	.99619	60
1	.08745	.08778	11.392	.99617	59
2	.08774	.08807	11.354	.99614	58
3	.08803	.08837	11.316	.99612	57
4	.08831	.08866	11.279	.99609	56
5	.08860	.08895	11.242	.99607	55
6	.08889	.08925	11.205	.99604	54
7	.08918	.08954	11.168	.99602	53
8	.08947	.08983	11.132	.99599	52
9	.08976	.09013	11.095	.99596	51
10	.09005	.09042	11.059	.99594	50
11	.09034	.09071	11.024	.99591	49
12	.09063	.09101	10.988	.99588	48
13	.09092	.09130	10.953	.99586	47
14	.09121	.09159	10.918	.99583	46
15	.09150	.09189	10.883	.99580	45
16	.09179	.09218	10.848	.99578	44
17	.09208	.09247	10.814	.99575	43
18	.09237	.09277	10.780	.99572	42
19	.09266	.09306	10.746	.99570	41
20	.09295	.09335	10.712	.99567	40
21	.09324	.09365	10.678	.99564	39
22	.09353	.09394	10.645	.99562	38
23	.09382	.09423	10.612	.99558	37
24	.09411	.09453	10.579	.99556	36
25	.09440	.09482	10.546	.99553	35
26	.09469	.09511	10.514	.99551	34
27	.09498	.09541	10.481	.99548	33
28	.09527	.09570	10.449	.99545	32
29	.09556	.09600	10.417	.99542	31
30	.09585	.09629	10.385	.99540	30
31	.09614	.09658	10.354	.99537	29
32	.09642	.09688	10.322	.99534	28
33	.09671	.09717	10.291	.99531	27
34	.09700	.09746	10.260	.99528	26
35	.09729	.09776	10.229	.99526	25
36	.09758	.09805	10.199	.99523	24
37	.09787	.09834	10.168	.99520	23
38	.09816	.09864	10.138	.99517	22
39	.09845	.09893	10.108	.99514	21
40	.09874	.09923	10.078	.99511	20
41	.09903	.09952	10.048	.99508	19
42	.09932	.09981	10.019	.99506	18
43	.09961	.10011	9.9893	.99503	17
44	.09990	.10040	9.9601	.99500	16
45	.10019	.10069	9.9310	.99497	15
46	.10048	.10099	9.9021	.99494	14
47	.10077	.10128	9.8734	.99491	13
48	.10106	.10158	9.8448	.99488	12
49	.10135	.10187	9.8164	.99485	11
50	.10164	.10216	9.7882	.99482	10
51	.10192	.10246	9.7601	.99479	9
52	.10221	.10275	9.7322	.99476	8
53	.10250	.10305	9.7044	.99473	7
54	.10279	.10334	9.6768	.99470	6
55	.10308	.10363	9.6493	.99467	5
56	.10337	.10393	9.6220	.99464	4
57	.10366	.10422	9.5949	.99461	3
58	.10395	.10452	9.5679	.99458	2
59	.10424	.10481	9.5411	.99455	1
60	.10453	.10510	9.5144	.99452	0
′	Cos	Cot	Tan	Sin	′

95° (275°)   **(264°) 84°**

NATURAL TRIGONOMETRIC FUNCTIONS

6° (186°)   **(353°) 173°**   **7° (187°)**   **(352°) 172°**

′	Sin	Tan	Cot	Cos	′
0	.10453	.10510	9.5144	.99452	60
1	.10482	.10540	9.4878	.99449	59
2	.10511	.10569	9.4614	.99446	58
3	.10540	.10599	9.4352	.99443	57
4	.10569	.10628	9.4090	.99440	56
5	.10597	.10657	9.3831	.99437	55
6	.10626	.10687	9.3572	.99434	54
7	.10655	.10716	9.3315	.99431	53
8	.10684	.10746	9.3060	.99428	52
9	.10713	.10775	9.2806	.99424	51
10	.10742	.10805	9.2553	.99421	50
11	.10771	.10834	9.2302	.99418	49
12	.10800	.10863	9.2052	.99415	48
13	.10829	.10893	9.1803	.99412	47
14	.10858	.10922	9.1555	.99409	46
15	.10887	.10952	9.1309	.99406	45
16	.10916	.10981	9.1065	.99402	44
17	.10945	.11011	9.0821	.99399	43
18	.10973	.11040	9.0579	.99396	42
19	.11002	.11070	9.0338	.99393	41
20	.11031	.11099	9.0098	.99390	40
21	.11060	.11128	8.9860	.99386	39
22	.11089	.11158	8.9623	.99383	38
23	.11118	.11187	8.9387	.99380	37
24	.11147	.11217	8.9152	.99377	36
25	.11176	.11246	8.8919	.99374	35
26	.11205	.11276	8.8686	.99370	34
27	.11234	.11305	8.8455	.99367	33
28	.11263	.11335	8.8225	.99364	32
29	.11291	.11364	8.7996	.99360	31
30	.11320	.11394	8.7769	.99357	30
31	.11349	.11423	8.7542	.99354	29
32	.11378	.11452	8.7317	.99351	28
33	.11407	.11482	8.7093	.99347	27
34	.11436	.11511	8.6870	.99344	26
35	.11465	.11541	8.6648	.99341	25
36	.11494	.11570	8.6427	.99337	24
37	.11523	.11600	8.6208	.99334	23
38	.11552	.11629	8.5989	.99331	22
39	.11580	.11659	8.5772	.99327	21
40	.11609	.11688	8.5555	.99324	20
41	.11638	.11718	8.5340	.99320	19
42	.11667	.11747	8.5126	.99317	18
43	.11696	.11777	8.4913	.99314	17
44	.11725	.11806	8.4701	.99310	16
45	.11754	.11836	8.4490	.99307	15
46	.11783	.11865	8.4280	.99303	14
47	.11812	.11895	8.4071	.99300	13
48	.11840	.11924	8.3863	.99297	12
49	.11869	.11954	8.3656	.99293	11
50	.11898	.11983	8.3450	.99290	10
51	.11927	.12013	8.3245	.99286	9
52	.11956	.12042	8.3041	.99283	8
53	.11985	.12072	8.2838	.99279	7
54	.12014	.12101	8.2636	.99276	6
55	.12043	.12131	8.2434	.99272	5
56	.12071	.12160	8.2234	.99269	4
57	.12100	.12190	8.2035	.99265	3
58	.12129	.12219	8.1837	.99262	2
59	.12158	.12249	8.1640	.99258	1
60	.12187	.12278	8.1443	.99255	0
′	Cos	Cot	Tan	Sin	′

96° (276°)   **(263°) 83°**

′	Sin	Tan	Cot	Cos	′
0	.12187	.12278	8.1443	.99255	60
1	.12216	.12308	8.1248	.99251	59
2	.12245	.12338	8.1054	.99248	58
3	.12274	.12367	8.0860	.99244	57
4	.12302	.12397	8.0667	.99240	56
5	.12331	.12426	8.0476	.99237	55
6	.12360	.12456	8.0285	.99233	54
7	.12389	.12485	8.0095	.99230	53
8	.12418	.12515	7.9906	.99226	52
9	.12447	.12544	7.9718	.99222	51
10	.12476	.12574	7.9530	.99219	50
11	.12504	.12603	7.9344	.99215	49
12	.12533	.12633	7.9158	.99211	48
13	.12562	.12662	7.8973	.99208	47
14	.12591	.12692	7.8789	.99204	46
15	.12620	.12722	7.8606	.99200	45
16	.12649	.12751	7.8424	.99197	44
17	.12678	.12781	7.8243	.99193	43
18	.12706	.12810	7.8062	.99189	42
19	.12735	.12840	7.7882	.99186	41
20	.12764	.12869	7.7704	.99182	40
21	.12793	.12899	7.7525	.99178	39
22	.12822	.12929	7.7348	.99175	38
23	.12851	.12958	7.7171	.99171	37
24	.12880	.12988	7.6996	.99167	36
25	.12908	.13017	7.6821	.99163	35
26	.12937	.13047	7.6647	.99160	34
27	.12966	.13076	7.6473	.99156	33
28	.12995	.13106	7.6301	.99152	32
29	.13024	.13136	7.6129	.99148	31
30	.13053	.13165	7.5958	.99144	30
31	.13081	.13195	7.5787	.99141	29
32	.13110	.13224	7.5618	.99137	28
33	.13139	.13254	7.5449	.99133	27
34	.13168	.13284	7.5281	.99129	26
35	.13197	.13313	7.5113	.99125	25
36	.13226	.13343	7.4947	.99122	24
37	.13254	.13372	7.4781	.99118	23
38	.13283	.13402	7.4615	.99114	22
39	.13312	.13432	7.4451	.99110	21
40	.13341	.13461	7.4287	.99106	20
41	.13370	.13491	7.4124	.99102	19
42	.13399	.13521	7.3962	.99098	18
43	.13427	.13550	7.3800	.99094	17
44	.13456	.13580	7.3639	.99091	16
45	.13485	.13609	7.3479	.99087	15
46	.13514	.13639	7.3319	.99083	14
47	.13543	.13669	7.3160	.99079	13
48	.13572	.13698	7.3002	.99075	12
49	.13600	.13728	7.2844	.99071	11
50	.13629	.13758	7.2687	.99067	10
51	.13658	.13787	7.2531	.99063	9
52	.13687	.13817	7.2375	.99059	8
53	.13716	.13846	7.2220	.99055	7
54	.13744	.13876	7.2066	.99051	6
55	.13773	.13906	7.1912	.99047	5
56	.13802	.13935	7.1759	.99043	4
57	.13831	.13965	7.1607	.99039	3
58	.13860	.13995	7.1455	.99035	2
59	.13889	.14024	7.1304	.99031	1
60	.13917	.14054	7.1154	.99027	0
′	Cos	Cot	Tan	Sin	′

97° (277°)   **(262°) 82°**

NATURAL TRIGONOMETRIC FUNCTIONS

8° (188°)   (351°) **171°**

′	Sin	Tan	Cot	Cos	′
0	.13917	.14054	7.1154	.99027	60
1	.13946	.14084	7.1004	.99023	59
2	.13975	.14113	7.0855	.99019	58
3	.14004	.14143	7.0706	.99015	57
4	.14033	.14173	7.0558	.99011	56
5	.14061	.14202	7.0410	.99006	55
6	.14090	.14232	7.0264	.99002	54
7	.14119	.14262	7.0117	.98998	53
8	.14148	.14291	6.9972	.98994	52
9	.14177	.14321	6.9827	.98990	51
10	.14205	.14351	6.9682	.98986	50
11	.14234	.14381	6.9538	.98982	49
12	.14263	.14410	6.9395	.98978	48
13	.14292	.14440	6.9252	.98973	47
14	.14320	.14470	6.9110	.98969	46
15	.14349	.14499	6.8969	.98965	45
16	.14378	.14529	6.8828	.98961	44
17	.14407	.14559	6.8687	.98957	43
18	.14436	.14588	6.8548	.98953	42
19	.14464	.14618	6.8408	.98948	41
20	.14493	.14648	6.8269	.98944	40
21	.14522	.14678	6.8131	.98940	39
22	.14551	.14707	6.7994	.98936	38
23	.14580	.14737	6.7856	.98931	37
24	.14608	.14767	6.7720	.98927	36
25	.14637	.14796	6.7584	.98923	35
26	.14666	.14826	6.7448	.98919	34
27	.14695	.14856	6.7313	.98914	33
28	.14723	.14886	6.7179	.98910	32
29	.14752	.14915	6.7045	.98906	31
30	.14781	.14945	6.6912	.98902	30
31	.14810	.14975	6.6779	.98897	29
32	.14838	.15005	6.6646	.98893	28
33	.14867	.15034	6.6514	.98889	27
34	.14896	.15064	6.6383	.98884	26
35	.14925	.15094	6.6252	.98880	25
36	.14954	.15124	6.6122	.98876	24
37	.14982	.15153	6.5992	.98871	23
38	.15011	.15183	6.5863	.98867	22
39	.15040	.15213	6.5734	.98863	21
40	.15069	.15243	6.5606	.98858	20
41	.15097	.15272	6.5478	.98854	19
42	.15126	.15302	6.5350	.98849	18
43	.15155	.15332	6.5223	.98845	17
44	.15184	.15362	6.5097	.98841	16
45	.15212	.15391	6.4971	.98836	15
46	.15241	.15421	6.4846	.98832	14
47	.15270	.15451	6.4721	.98827	13
48	.15299	.15481	6.4596	.98823	12
49	.15327	.15511	6.4472	.98818	11
50	.15356	.15540	6.4348	.98814	10
51	.15385	.15570	6.4225	.98809	9
52	.15414	.15600	6.4103	.98805	8
53	.15442	.15630	6.3980	.98800	7
54	.15471	.15660	6.3859	.98796	6
55	.15500	.15689	6.3737	.98791	5
56	.15529	.15719	6.3617	.98787	4
57	.15557	.15749	6.3496	.98782	3
58	.15586	.15779	6.3376	.98778	2
59	.15615	.15809	6.3257	.98773	1
60	.15643	.15838	6.3138	.98769	0
′	Cos	Cot	Tan	Sin	′

9° (189°)   (350°) **170°**

′	Sin	Tan	Cot	Cos	′
0	.15643	.15838	6.3138	.98769	60
1	.15672	.15868	6.3019	.98764	59
2	.15701	.15898	6.2901	.98760	58
3	.15730	.15928	6.2783	.98755	57
4	.15758	.15958	6.2666	.98751	56
5	.15787	.15988	6.2549	.98746	55
6	.15816	.16017	6.2432	.98741	54
7	.15845	.16047	6.2316	.98737	53
8	.15873	.16077	6.2200	.98732	52
9	.15902	.16107	6.2085	.98728	51
10	.15931	.16137	6.1970	.98723	50
11	.15959	.16167	6.1856	.98718	49
12	.15988	.16196	6.1742	.98714	48
13	.16017	.16226	6.1628	.98709	47
14	.16046	.16256	6.1515	.98704	46
15	.16074	.16286	6.1402	.98700	45
16	.16103	.16316	6.1290	.98695	44
17	.16132	.16346	6.1178	.98690	43
18	.16160	.16376	6.1066	.98686	42
19	.16189	.16405	6.0955	.98681	41
20	.16218	.16435	6.0844	.98676	40
21	.16246	.16465	6.0734	.98671	39
22	.16275	.16495	6.0624	.98667	38
23	.16304	.16525	6.0514	.98662	37
24	.16333	.16555	6.0405	.98657	36
25	.16361	.16585	6.0296	.98652	35
26	.16390	.16615	6.0188	.98648	34
27	.16419	.16645	6.0080	.98643	33
28	.16447	.16674	5.9972	.98638	32
29	.16476	.16704	5.9865	.98633	31
30	.16505	.16734	5.9758	.98629	30
31	.16533	.16764	5.9651	.98624	29
32	.16562	.16794	5.9545	.98619	28
33	.16591	.16824	5.9439	.98614	27
34	.16620	.16854	5.9333	.98609	26
35	.16648	.16884	5.9228	.98604	25
36	.16677	.16914	5.9124	.98600	24
37	.16706	.16944	5.9019	.98595	23
38	.16734	.16974	5.8915	.98590	22
39	.16763	.17004	5.8811	.98585	21
40	.16792	.17033	5.8708	.98580	20
41	.16820	.17063	5.8605	.98575	19
42	.16849	.17093	5.8502	.98570	18
43	.16878	.17123	5.8400	.98565	17
44	.16906	.17153	5.8298	.98561	16
45	.16935	.17183	5.8197	.98556	15
46	.16964	.17213	5.8095	.98551	14
47	.16992	.17243	5.7994	.98546	13
48	.17021	.17273	5.7894	.98541	12
49	.17050	.17303	5.7794	.98536	11
50	.17078	.17333	5.7694	.98531	10
51	.17107	.17363	5.7594	.98526	9
52	.17136	.17393	5.7495	.98521	8
53	.17164	.17423	5.7396	.98516	7
54	.17193	.17453	5.7297	.98511	6
55	.17222	.17483	5.7199	.98506	5
56	.17250	.17513	5.7101	.98501	4
57	.17279	.17543	5.7004	.98496	3
58	.17308	.17573	5.6906	.98491	2
59	.17336	.17603	5.6809	.98486	1
60	.17365	.17633	5.6713	.98481	0
′	Cos	Cot	Tan	Sin	′

NATURAL TRIGONOMETRIC FUNCTIONS

10° (190°)   **(349°) 169°**

'	Sin	Tan	Cot	Cos	'
0	.17365	.17633	5.6713	.98481	60
1	.17393	.17663	5.6617	.98476	59
2	.17422	.17693	5.6521	.98471	58
3	.17451	.17723	5.6425	.98466	57
4	.17479	.17753	5.6329	.98461	56
5	.17508	.17783	5.6234	.98455	55
6	.17537	.17813	5.6140	.98450	54
7	.17565	.17843	5.6045	.98445	53
8	.17594	.17873	5.5951	.98440	52
9	.17623	.17903	5.5857	.98435	51
10	.17651	.17933	5.5764	.98430	50
11	.17680	.17963	5.5671	.98425	49
12	.17708	.17993	5.5578	.98420	48
13	.17737	.18023	5.5485	.98414	47
14	.17766	.18053	5.5393	.98409	46
15	.17794	.18083	5.5301	.98404	45
16	.17823	.18113	5.5209	.98399	44
17	.17852	.18143	5.5118	.98394	43
18	.17880	.18173	5.5026	.98389	42
19	.17909	.18203	5.4936	.98383	41
20	.17937	.18233	5.4845	.98378	40
21	.17966	.18263	5.4755	.98373	39
22	.17995	.18293	5.4665	.98368	38
23	.18023	.18323	5.4575	.98362	37
24	.18052	.18353	5.4486	.98357	36
25	.18081	.18384	5.4397	.98352	35
26	.18109	.18414	5.4308	.98347	34
27	.18138	.18444	5.4219	.98341	33
28	.18166	.18474	5.4131	.98336	32
29	.18195	.18504	5.4043	.98331	31
30	.18224	.18534	5.3955	.98325	30
31	.18252	.18564	5.3868	.98320	29
32	.18281	.18594	5.3781	.98315	28
33	.18309	.18624	5.3694	.98310	27
34	.18338	.18654	5.3607	.98304	26
35	.18367	.18684	5.3521	.98299	25
36	.18395	.18714	5.3435	.98294	24
37	.18424	.18745	5.3349	.98288	23
38	.18452	.18775	5.3263	.98283	22
39	.18481	.18805	5.3178	.98277	21
40	.18509	.18835	5.3093	.98272	20
41	.18538	.18865	5.3008	.98267	19
42	.18567	.18895	5.2924	.98261	18
43	.18595	.18925	5.2839	.98256	17
44	.18624	.18955	5.2755	.98250	16
45	.18652	.18986	5.2672	.98245	15
46	.18681	.19016	5.2588	.98240	14
47	.18710	.19046	5.2505	.98234	13
48	.18738	.19076	5.2422	.98229	12
49	.18767	.19106	5.2339	.98223	11
50	.18795	.19136	5.2257	.98218	10
51	.18824	.19166	5.2174	.98212	9
52	.18852	.19197	5.2092	.98207	8
53	.18881	.19227	5.2011	.98201	7
54	.18910	.19257	5.1929	.98196	6
55	.18938	.19287	5.1848	.98190	5
56	.18967	.19317	5.1767	.98185	4
57	.18995	.19347	5.1686	.98179	3
58	.19024	.19378	5.1606	.98174	2
59	.19052	.19408	5.1526	.98168	1
60	.19081	.19438	5.1446	.98163	0
'	Cos	Cot	Tan	Sin	'

11° (191°)   **(348°) 168°**

'	Sin	Tan	Cot	Cos	'
0	.19081	.19438	5.1446	.98163	60
1	.19109	.19468	5.1366	.98157	59
2	.19138	.19498	5.1286	.98152	58
3	.19167	.19529	5.1207	.98146	57
4	.19195	.19559	5.1128	.98140	56
5	.19224	.19589	5.1049	.98135	55
6	.19252	.19619	5.0970	.98129	54
7	.19281	.19649	5.0892	.98124	53
8	.19309	.19680	5.0814	.98118	52
9	.19338	.19710	5.0736	.98112	51
10	.19366	.19740	5.0658	.98107	50
11	.19395	.19770	5.0581	.98101	49
12	.19423	.19801	5.0504	.98096	48
13	.19452	.19831	5.0427	.98090	47
14	.19481	.19861	5.0350	.98084	46
15	.19509	.19891	5.0273	.98079	45
16	.19538	.19921	5.0197	.98073	44
17	.19566	.19952	5.0121	.98067	43
18	.19595	.19982	5.0045	.98061	42
19	.19623	.20012	4.9969	.98056	41
20	.19652	.20042	4.9894	.98050	40
21	.19680	.20073	4.9819	.98044	39
22	.19709	.20103	4.9744	.98039	38
23	.19737	.20133	4.9669	.98033	37
24	.19766	.20164	4.9594	.98027	36
25	.19794	.20194	4.9520	.98021	35
26	.19823	.20224	4.9446	.98016	34
27	.19851	.20254	4.9372	.98010	33
28	.19880	.20285	4.9298	.98004	32
29	.19908	.20315	4.9225	.97998	31
30	.19937	.20345	4.9152	.97992	30
31	.19965	.20376	4.9078	.97987	29
32	.19994	.20406	4.9006	.97981	28
33	.20022	.20436	4.8933	.97975	27
34	.20051	.20466	4.8860	.97969	26
35	.20079	.20497	4.8788	.97963	25
36	.20108	.20527	4.8716	.97958	24
37	.20136	.20557	4.8644	.97952	23
38	.20165	.20588	4.8573	.97946	22
39	.20193	.20618	4.8501	.97940	21
40	.20222	.20648	4.8430	.97934	20
41	.20250	.20679	4.8359	.97928	19
42	.20279	.20709	4.8288	.97922	18
43	.20307	.20739	4.8218	.97916	17
44	.20336	.20770	4.8147	.97910	16
45	.20364	.20800	4.8077	.97905	15
46	.20393	.20830	4.8007	.97899	14
47	.20421	.20861	4.7937	.97893	13
48	.20450	.20891	4.7867	.97887	12
49	.20478	.20921	4.7798	.97881	11
50	.20507	.20952	4.7729	.97875	10
51	.20535	.20982	4.7659	.97869	9
52	.20563	.21013	4.7591	.97863	8
53	.20592	.21043	4.7522	.97857	7
54	.20620	.21073	4.7453	.97851	6
55	.20649	.21104	4.7385	.97845	5
56	.20677	.21134	4.7317	.97839	4
57	.20706	.21164	4.7249	.97833	3
58	.20734	.21195	4.7181	.97827	2
59	.20763	.21225	4.7114	.97821	1
60	.20791	.21256	4.7046	.97815	0
'	Cos	Cot	Tan	Sin	'

NATURAL TRIGONOMETRIC FUNCTIONS

12° (192°)   **(347°) 167°**   **13° (193°)**   **(346°) 166°**

′	Sin	Tan	Cot	Cos	′		′	Sin	Tan	Cot	Cos	′
0	.20791	.21256	4.7046	.97815	60		0	.22495	.23087	4.3315	.97437	60
1	.20820	.21286	4.6979	.97809	59		1	.22523	.23117	4.3257	.97430	59
2	.20848	.21316	4.6912	.97803	58		2	.22552	.23148	4.3200	.97424	58
3	.20877	.21347	4.6845	.97797	57		3	.22580	.23179	4.3143	.97417	57
4	.20905	.21377	4.6779	.97791	56		4	.22608	.23209	4.3086	.97411	56
5	.20933	.21408	4.6712	.97784	55		5	.22637	.23240	4.3029	.97404	55
6	.20962	.21438	4.6646	.97778	54		6	.22665	.23271	4.2972	.97398	54
7	.20990	.21469	4.6580	.97772	53		7	.22693	.23301	4.2916	.97391	53
8	.21019	.21499	4.6514	.97766	52		8	.22722	.23332	4.2859	.97384	52
9	.21047	.21529	4.6448	.97760	51		9	.22750	.23363	4.2803	.97378	51
10	.21076	.21560	4.6382	.97754	50		10	.22778	.23393	4.2747	.97371	50
11	.21104	.21590	4.6317	.97748	49		11	.22807	.23424	4.2691	.97365	49
12	.21132	.21621	4.6252	.97742	48		12	.22835	.23455	4.2635	.97358	48
13	.21161	.21651	4.6187	.97735	47		13	.22863	.23485	4.2580	.97351	47
14	.21189	.21682	4.6122	.97729	46		14	.22892	.23516	4.2524	.97345	46
15	.21218	.21712	4.6057	.97723	45		15	.22920	.23547	4.2468	.97338	45
16	.21246	.21743	4.5993	.97717	44		16	.22948	.23578	4.2413	.97331	44
17	.21275	.21773	4.5928	.97711	43		17	.22977	.23608	4.2358	.97325	43
18	.21303	.21804	4.5864	.97705	42		18	.23005	.23639	4.2303	.97318	42
19	.21331	.21834	4.5800	.97698	41		19	.23033	.23670	4.2248	.97311	41
20	.21360	.21864	4.5736	.97692	40		20	.23062	.23700	4.2193	.97304	40
21	.21388	.21895	4.5673	.97686	39		21	.23090	.23731	4.2139	.97298	39
22	.21417	.21925	4.5609	.97680	38		22	.23118	.23762	4.2084	.97291	38
23	.21445	.21956	4.5546	.97673	37		23	.23146	.23793	4.2030	.97284	37
24	.21474	.21986	4.5483	.97667	36		24	.23175	.23823	4.1976	.97278	36
25	.21502	.22017	4.5420	.97661	35		25	.23203	.23854	4.1922	.97271	35
26	.21530	.22047	4.5357	.97655	34		26	.23231	.23885	4.1868	.97264	34
27	.21559	.22078	4.5294	.97648	33		27	.23260	.23916	4.1814	.97257	33
28	.21587	.22108	4.5232	.97642	32		28	.23288	.23946	4.1760	.97251	32
29	.21616	.22139	4.5169	.97636	31		29	.23316	.23977	4.1706	.97244	31
30	.21644	.22169	4.5107	.97630	30		30	.23345	.24008	4.1653	.97237	30
31	.21672	.22200	4.5045	.97623	29		31	.23373	.24039	4.1600	.97230	29
32	.21701	.22231	4.4983	.97617	28		32	.23401	.24069	4.1547	.97223	28
33	.21729	.22261	4.4922	.97611	27		33	.23429	.24100	4.1493	.97217	27
34	.21758	.22292	4.4860	.97604	26		34	.23458	.24131	4.1441	.97210	26
35	.21786	.22322	4.4799	.97598	25		35	.23486	.24162	4.1388	.97203	25
36	.21814	.22353	4.4737	.97592	24		36	.23514	.24193	4.1335	.97196	24
37	.21843	.22383	4.4676	.97585	23		37	.23542	.24223	4.1282	.97189	23
38	.21871	.22414	4.4615	.97579	22		38	.23571	.24254	4.1230	.97182	22
39	.21899	.22444	4.4555	.97573	21		39	.23599	.24285	4.1178	.97176	21
40	.21928	.22475	4.4494	.97566	20		40	.23627	.24316	4.1126	.97169	20
41	.21956	.22505	4.4434	.97560	19		41	.23656	.24347	4.1074	.97162	19
42	.21985	.22536	4.4373	.97553	18		42	.23684	.24377	4.1022	.97155	18
43	.22013	.22567	4.4313	.97547	17		43	.23712	.24408	4.0970	.97148	17
44	.22041	.22597	4.4253	.97541	16		44	.23740	.24439	4.0918	.97141	16
45	.22070	.22628	4.4194	.97534	15		45	.23769	.24470	4.0867	.97134	15
46	.22098	.22658	4.4134	.97528	14		46	.23797	.24501	4.0815	.97127	14
47	.22126	.22689	4.4075	.97521	13		47	.23825	.24532	4.0764	.97120	13
48	.22155	.22719	4.4015	.97515	12		48	.23853	.24562	4.0713	.97113	12
49	.22183	.22750	4.3956	.97508	11		49	.23882	.24593	4.0662	.97106	11
50	.22212	.22781	4.3897	.97502	10		50	.23910	.24624	4.0611	.97100	10
51	.22240	.22811	4.3838	.97496	9		51	.23938	.24655	4.0560	.97093	9
52	.22268	.22842	4.3779	.97489	8		52	.23966	.24686	4.0509	.97086	8
53	.22297	.22872	4.3721	.97483	7		53	.23995	.24717	4.0459	.97079	7
54	.22325	.22903	4.3662	.97476	6		54	.24023	.24747	4.0408	.97072	6
55	.22353	.22934	4.3604	.97470	5		55	.24051	.24778	4.0358	.97065	5
56	.22382	.22964	4.3546	.97463	4		56	.24079	.24809	4.0308	.97058	4
57	.22410	.22995	4.3488	.97457	3		57	.24108	.24840	4.0257	.97051	3
58	.22438	.23026	4.3430	.97450	2		58	.24136	.24871	4.0207	.97044	2
59	.22467	.23056	4.3372	.97444	1		59	.24164	.24902	4.0158	.97037	1
60	.22495	.23087	4.3315	.97437	0		60	.24192	.24933	4.0108	.97030	0
′	Cos	Cot	Tan	Sin	′		′	Cos	Cot	Tan	Sin	′

102° (282°)   **(257°) 77°**   **103° (283°)**   **(256°) 76°**

NATURAL TRIGONOMETRIC FUNCTIONS

14° (194°)   (345°) **165°**   **15° (195°)**   (344°) **164°**

′	Sin	Tan	Cot	Cos	′		′	Sin	Tan	Cot	Cos	′
0	.24192	.24933	4.0108	.97030	60		0	.25882	.26795	3.7321	.96593	60
1	.24220	.24964	4.0058	.97023	59		1	.25910	.26826	3.7277	.96585	59
2	.24249	.24995	4.0009	.97015	58		2	.25938	.26857	3.7234	.96578	58
3	.24277	.25026	3.9959	.97008	57		3	.25966	.26888	3.7191	.96570	57
4	.24305	.25056	3.9910	.97001	56		4	.25994	.26920	3.7148	.96562	56
5	.24333	.25087	3.9861	.96994	55		5	.26022	.26951	3.7105	.96555	55
6	.24362	.25118	3.9812	.96987	54		6	.26050	.26982	3.7062	.96547	54
7	.24390	.25149	3.9763	.96980	53		7	.26079	.27013	3.7019	.96540	53
8	.24418	.25180	3.9714	.96973	52		8	.26107	.27044	3.6976	.96532	52
9	.24446	.25211	3.9665	.96966	51		9	.26135	.27076	3.6933	.96524	51
10	.24474	.25242	3.9617	.96959	50		10	.26163	.27107	3.6891	.96517	50
11	.24503	.25273	3.9568	.96952	49		11	.26191	.27138	3.6848	.96509	49
12	.24531	.25304	3.9520	.96945	48		12	.26219	.27169	3.6806	.96502	48
13	.24559	.25335	3.9471	.96937	47		13	.26247	.27201	3.6764	.96494	47
14	.24587	.25366	3.9423	.96930	46		14	.26275	.27232	3.6722	.96486	46
15	.24615	.25397	3.9375	.96923	45		15	.26303	.27263	3.6680	.96479	45
16	.24644	.25428	3.9327	.96916	44		16	.26331	.27294	3.6638	.96471	44
17	.24672	.25459	3.9279	.96909	43		17	.26359	.27326	3.6596	.96463	43
18	.24700	.25490	3.9232	.96902	42		18	.26387	.27357	3.6554	.96456	42
19	.24728	.25521	3.9184	.96894	41		19	.26415	.27388	3.6512	.96448	41
20	.24756	.25552	3.9136	.96887	40		20	.26443	.27419	3.6470	.96440	40
21	.24784	.25583	3.9089	.96880	39		21	.26471	.27451	3.6429	.96433	39
22	.24813	.25614	3.9042	.96873	38		22	.26500	.27482	3.6387	.96425	38
23	.24841	.25645	3.8995	.96866	37		23	.26528	.27513	3.6346	.96417	37
24	.24869	.25676	3.8947	.96858	36		24	.26556	.27545	3.6305	.96410	36
25	.24897	.25707	3.8900	.96851	35		25	.26584	.27576	3.6264	.96402	35
26	.24925	.25738	3.8854	.96844	34		26	.26612	.27607	3.6222	.96394	34
27	.24954	.25769	3.8807	.96837	33		27	.26640	.27638	3.6181	.96386	33
28	.24982	.25800	3.8760	.96829	32		28	.26668	.27670	3.6140	.96379	32
29	.25010	.25831	3.8714	.96822	31		29	.26696	.27701	3.6100	.96371	31
30	.25038	.25862	3.8667	.96815	30		30	.26724	.27732	3.6059	.96363	30
31	.25066	.25893	3.8621	.96807	29		31	.26752	.27764	3.6018	.96355	29
32	.25094	.25924	3.8575	.96800	28		32	.26780	.27795	3.5978	.96347	28
33	.25122	.25955	3.8528	.96793	27		33	.26808	.27826	3.5937	.96340	27
34	.25151	.25986	3.8482	.96786	26		34	.26836	.27858	3.5897	.96332	26
35	.25179	.26017	3.8436	.96778	25		35	.26864	.27889	3.5856	.96324	25
36	.25207	.26048	3.8391	.96771	24		36	.26892	.27921	3.5816	.96316	24
37	.25235	.26079	3.8345	.96764	23		37	.26920	.27952	3.5776	.96308	23
38	.25263	.26110	3.8299	.96756	22		38	.26948	.27983	3.5736	.96301	22
39	.25291	.26141	3.8254	.96749	21		39	.26976	.28015	3.5696	.96293	21
40	.25320	.26172	3.8208	.96742	20		40	.27004	.28046	3.5656	.96285	20
41	.25348	.26203	3.8163	.96734	19		41	.27032	.28077	3.5616	.96277	19
42	.25376	.26235	3.8118	.96727	18		42	.27060	.28109	3.5576	.96269	18
43	.25404	.26266	3.8073	.96719	17		43	.27088	.28140	3.5536	.96261	17
44	.25432	.26297	3.8028	.96712	16		44	.27116	.28172	3.5497	.96253	16
45	.25460	.26328	3.7983	.96705	15		45	.27144	.28203	3.5457	.96246	15
46	.25488	.26359	3.7938	.96697	14		46	.27172	.28234	3.5418	.96238	14
47	.25516	.26390	3.7893	.96690	13		47	.27200	.28266	3.5379	.96230	13
48	.25545	.26421	3.7848	.96682	12		48	.27228	.28297	3.5339	.96222	12
49	.25573	.26452	3.7804	.96675	11		49	.27256	.28329	3.5300	.96214	11
50	.25601	.26483	3.7760	.96667	10		50	.27284	.28360	3.5261	.96206	10
51	.25629	.26515	3.7715	.96660	9		51	.27312	.28391	3.5222	.96198	9
52	.25657	.26546	3.7671	.96653	8		52	.27340	.28423	3.5183	.96190	8
53	.25685	.26577	3.7627	.96645	7		53	.27368	.28454	3.5144	.96182	7
54	.25713	.26608	3.7583	.96638	6		54	.27396	.28486	3.5105	.96174	6
55	.25741	.26639	3.7539	.96630	5		55	.27424	.28517	3.5067	.96166	5
56	.25769	.26670	3.7495	.96623	4		56	.27452	.28549	3.5028	.96158	4
57	.25798	.26701	3.7451	.96615	3		57	.27480	.28580	3.4989	.96150	3
58	.25826	.26733	3.7408	.96608	2		58	.27508	.28612	3.4951	.96142	2
59	.25854	.26764	3.7364	.96600	1		59	.27536	.28643	3.4912	.96134	1
60	.25882	.26795	3.7321	.96593	0		60	.27564	.28675	3.4874	.96126	0
′	Cos	Cot	Tan	Sin	′		′	Cos	Cot	Tan	Sin	′

104° (284°)   (255°) **75°**   **105° (285°)**   (254°) **74°**

NATURAL TRIGONOMETRIC FUNCTIONS

16° (196°)   (343°) **163°**   **17° (197°)**   (342°) **162°**

′	Sin	Tan	Cot	Cos	′		′	Sin	Tan	Cot	Cos	′
0	.27564	.28675	3.4874	.96126	60		0	.29237	.30573	3.2709	.95630	60
1	.27592	.28706	3.4836	.96118	59		1	.29265	.30605	3.2675	.95622	59
2	.27620	.28738	3.4798	.96110	58		2	.29293	.30637	3.2641	.95613	58
3	.27648	.28769	3.4760	.96102	57		3	.29321	.30669	3.2607	.95605	57
4	.27676	.28801	3.4722	.96094	56		4	.29348	.30700	3.2573	.95596	56
5	.27704	.28832	3.4684	.96086	55		5	.29376	.30732	3.2539	.95588	55
6	.27731	.28864	3.4646	.96078	54		6	.29404	.30764	3.2506	.95579	54
7	.27759	.28895	3.4608	.96070	53		7	.29432	.30796	3.2472	.95571	53
8	.27787	.28927	3.4570	.96062	52		8	.29460	.30828	3.2438	.95562	52
9	.27815	.28958	3.4533	.96054	51		9	.29487	.30860	3.2405	.95554	51
10	.27843	.28990	3.4495	.96046	50		10	.29515	.30891	3.2371	.95545	50
11	.27871	.29021	3.4458	.96037	49		11	.29543	.30923	3.2338	.95536	49
12	.27900	.29053	3.4420	.96029	48		12	.29571	.30955	3.2305	.95528	48
13	.27927	.29084	3.4383	.96021	47		13	.29599	.30987	3.2272	.95519	47
14	.27955	.29116	3.4346	.96013	46		14	.29626	.31019	3.2238	.95511	46
15	.27983	.29147	3.4308	.96005	45		15	.29654	.31051	3.2205	.95502	45
16	.28011	.29179	3.4271	.95997	44		16	.29682	.31083	3.2172	.95493	44
17	.28039	.29210	3.4234	.95989	43		17	.29710	.31115	3.2139	.95485	43
18	.28067	.29242	3.4197	.95981	42		18	.29737	.31147	3.2106	.95476	42
19	.28095	.29274	3.4160	.95972	41		19	.29765	.31178	3.2073	.95467	41
20	.28123	.29305	3.4124	.95964	40		20	.29793	.31210	3.2041	.95459	40
21	.28150	.29337	3.4087	.95956	39		21	.29821	.31242	3.2008	.95450	39
22	.28178	.29368	3.4050	.95948	38		22	.29849	.31274	3.1975	.95441	38
23	.28206	.29400	3.4014	.95940	37		23	.29876	.31306	3.1943	.95433	37
24	.28234	.29432	3.3977	.95931	36		24	.29904	.31338	3.1910	.95424	36
25	.28262	.29463	3.3941	.95923	35		25	.29932	.31370	3.1878	.95415	35
26	.28290	.29495	3.3904	.95915	34		26	.29960	.31402	3.1845	.95407	34
27	.28318	.29526	3.3868	.95907	33		27	.29987	.31434	3.1813	.95398	33
28	.28346	.29558	3.3832	.95898	32		28	.30015	.31466	3.1780	.95389	32
29	.28374	.29590	3.3796	.95890	31		29	.30043	.31498	3.1748	.95380	31
30	.28402	.29621	3.3759	.95882	30		30	.30071	.31530	3.1716	.95372	30
31	.28429	.29653	3.3723	.95874	29		31	.30098	.31562	3.1684	.95363	29
32	.28457	.29685	3.3687	.95865	28		32	.30126	.31594	3.1652	.95354	28
33	.28485	.29716	3.3652	.95857	27		33	.30154	.31626	3.1620	.95345	27
34	.28513	.29748	3.3616	.95849	26		34	.30182	.31658	3.1588	.95337	26
35	.28541	.29780	3.3580	.95841	25		35	.30209	.31690	3.1556	.95328	25
36	.28569	.29811	3.3544	.95832	24		36	.30237	.31722	3.1524	.95319	24
37	.28597	.29843	3.3509	.95824	23		37	.30265	.31754	3.1492	.95310	23
38	.28625	.29875	3.3473	.95816	22		38	.30292	.31786	3.1460	.95301	22
39	.28652	.29906	3.3438	.95807	21		39	.30320	.31818	3.1429	.95293	21
40	.28680	.29938	3.3402	.95799	20		40	.30348	.31850	3.1397	.95284	20
41	.28708	.29970	3.3367	.95791	19		41	.30376	.31882	3.1366	.95275	19
42	.28736	.30001	3.3332	.95782	18		42	.30403	.31914	3.1334	.95266	18
43	.28764	.30033	3.3297	.95774	17		43	.30431	.31946	3.1303	.95257	17
44	.28792	.30065	3.3261	.95766	16		44	.30459	.31978	3.1271	.95248	16
45	.28820	.30097	3.3226	.95757	15		45	.30486	.32010	3.1240	.95240	15
46	.28847	.30128	3.3191	.95749	14		46	.30514	.32042	3.1209	.95231	14
47	.28875	.30160	3.3156	.95740	13		47	.30542	.32074	3.1178	.95222	13
48	.28903	.30192	3.3122	.95732	12		48	.30570	.32106	3.1146	.95213	12
49	.28931	.30224	3.3087	.95724	11		49	.30597	.32139	3.1115	.95204	11
50	.28959	.30255	3.3052	.95715	10		50	.30625	.32171	3.1084	.95195	10
51	.28987	.30287	3.3017	.95707	9		51	.30653	.32203	3.1053	.95186	9
52	.29015	.30319	3.2983	.95698	8		52	.30680	.32235	3.1022	.95177	8
53	.29042	.30351	3.2948	.95690	7		53	.30708	.32267	3.0991	.95168	7
54	.29070	.30382	3.2914	.95681	6		54	.30736	.32299	3.0961	.95159	6
55	.29098	.30414	3.2879	.95673	5		55	.30763	.32331	3.0930	.95150	5
56	.29126	.30446	3.2845	.95664	4		56	.30791	.32363	3.0899	.95142	4
57	.29154	.30478	3.2811	.95656	3		57	.30819	.32396	3.0868	.95133	3
58	.29182	.30509	3.2777	.95647	2		58	.30846	.32428	3.0838	.95124	2
59	.29209	.30541	3.2743	.95639	1		59	.30874	.32460	3.0807	.95115	1
60	.29237	.30573	3.2709	.95630	0		60	.30902	.32492	3.0777	.95106	0
′	Cos	Cot	Tan	Sin	′		′	Cos	Cot	Tan	Sin	′

106° (286°)   (253°) **73°**   **107° (287°)**   (252°) **72°**

NATURAL TRIGONOMETRIC FUNCTIONS

18° (198°) (341°) **161°**   **19° (199°)** (340°) **160°**

′	Sin	Tan	Cot	Cos	′		′	Sin	Tan	Cot	Cos	′
0	.30902	.32492	3.0777	.95106	60		0	.32557	.34433	2.9042	.94552	60
1	.30929	.32524	3.0746	.95097	59		1	.32584	.34465	2.9015	.94542	59
2	.30957	.32556	3.0716	.95088	58		2	.32612	.34498	2.8987	.94533	58
3	.30985	.32588	3.0686	.95079	57		3	.32639	.34530	2.8960	.94523	57
4	.31012	.32621	3.0655	.95070	56		4	.32667	.34563	2.8933	.94514	56
5	.31040	.32653	3.0625	.95061	55		5	.32694	.34596	2.8905	.94504	55
6	.31068	.32685	3.0595	.95052	54		6	.32722	.34628	2.8878	.94495	54
7	.31095	.32717	3.0565	.95043	53		7	.32749	.34661	2.8851	.94485	53
8	.31123	.32749	3.0535	.95033	52		8	.32777	.34693	2.8824	.94476	52
9	.31151	.32782	3.0505	.95024	51		9	.32804	.34726	2.8797	.94466	51
10	.31178	.32814	3.0475	.95015	50		10	.32832	.34758	2.8770	.94457	50
11	.31206	.32846	3.0445	.95006	49		11	.32859	.34791	2.8743	.94447	49
12	.31233	.32878	3.0415	.94997	48		12	.32887	.34824	2.8716	.94438	48
13	.31261	.32911	3.0385	.94988	47		13	.32914	.34856	2.8689	.94428	47
14	.31289	.32943	3.0356	.94979	46		14	.32942	.34889	2.8662	.94418	46
15	.31316	.32975	3.0326	.94970	45		15	.32969	.34922	2.8636	.94409	45
16	.31344	.33007	3.0296	.94961	44		16	.32997	.34954	2.8609	.94399	44
17	.31372	.33040	3.0267	.94952	43		17	.33024	.34987	2.8582	.94390	43
18	.31399	.33072	3.0237	.94943	42		18	.33051	.35020	2.8556	.94380	42
19	.31427	.33104	3.0208	.94933	41		19	.33079	.35052	2.8529	.94370	41
20	.31454	.33136	3.0178	.94924	40		20	.33106	.35085	2.8502	.94361	40
21	.31482	.33169	3.0149	.94915	39		21	.33134	.35118	2.8476	.94351	39
22	.31510	.33201	3.0120	.94906	38		22	.33161	.35150	2.8449	.94342	38
23	.31537	.33233	3.0090	.94897	37		23	.33189	.35183	2.8423	.94332	37
24	.31565	.33266	3.0061	.94888	36		24	.33216	.35216	2.8397	.94322	36
25	.31593	.33298	3.0032	.94878	35		25	.33244	.35248	2.8370	.94313	35
26	.31620	.33330	3.0003	.94869	34		26	.33271	.35281	2.8344	.94303	34
27	.31648	.33363	2.9974	.94860	33		27	.33298	.35314	2.8318	.94293	33
28	.31675	.33395	2.9945	.94851	32		28	.33326	.35346	2.8291	.94284	32
29	.31703	.33427	2.9916	.94842	31		29	.33353	.35379	2.8265	.94274	31
30	.31730	.33460	2.9887	.94832	30		30	.33381	.35412	2.8239	.94264	30
31	.31758	.33492	2.9858	.94823	29		31	.33408	.35445	2.8213	.94254	29
32	.31786	.33524	2.9829	.94814	28		32	.33436	.35477	2.8187	.94245	28
33	.31813	.33557	2.9800	.94805	27		33	.33463	.35510	2.8161	.94235	27
34	.31841	.33589	2.9772	.94795	26		34	.33490	.35543	2.8135	.94225	26
35	.31868	.33621	2.9743	.94786	25		35	.33518	.35576	2.8109	.94215	25
36	.31896	.33654	2.9714	.94777	24		36	.33545	.35608	2.8083	.94206	24
37	.31923	.33686	2.9686	.94768	23		37	.33573	.35641	2.8057	.94196	23
38	.31951	.33718	2.9657	.94758	22		38	.33600	.35674	2.8032	.94186	22
39	.31979	.33751	2.9629	.94749	21		39	.33627	.35707	2.8006	.94176	21
40	.32006	.33783	2.9600	.94740	20		40	.33655	.35740	2.7980	.94167	20
41	.32034	.33816	2.9572	.94730	19		41	.33682	.35772	2.7955	.94157	19
42	.32061	.33848	2.9544	.94721	18		42	.33710	.35805	2.7929	.94147	18
43	.32089	.33881	2.9515	.94712	17		43	.33737	.35838	2.7903	.94137	17
44	.32116	.33913	2.9487	.94702	16		44	.33764	.35871	2.7878	.94127	16
45	.32144	.33945	2.9459	.94693	15		45	.33792	.35904	2.7852	.94118	15
46	.32171	.33978	2.9431	.94684	14		46	.33819	.35937	2.7827	.94108	14
47	.32199	.34010	2.9403	.94674	13		47	.33846	.35969	2.7801	.94098	13
48	.32227	.34043	2.9375	.94665	12		48	.33874	.36002	2.7776	.94088	12
49	.32254	.34075	2.9347	.94656	11		49	.33901	.36035	2.7751	.94078	11
50	.32282	.34108	2.9319	.94646	10		50	.33929	.36068	2.7725	.94068	10
51	.32309	.34140	2.9291	.94637	9		51	.33956	.36101	2.7700	.94058	9
52	.32337	.34173	2.9263	.94627	8		52	.33983	.36134	2.7675	.94049	8
53	.32364	.34205	2.9235	.94618	7		53	.34011	.36167	2.7650	.94039	7
54	.32392	.34238	2.9208	.94609	6		54	.34038	.36199	2.7625	.94029	6
55	.32419	.34270	2.9180	.94599	5		55	.34065	.36232	2.7600	.94019	5
56	.32447	.34303	2.9152	.94590	4		56	.34093	.36265	2.7575	.94009	4
57	.32474	.34335	2.9125	.94580	3		57	.34120	.36298	2.7550	.93999	3
58	.32502	.34368	2.9097	.94571	2		58	.34147	.36331	2.7525	.93989	2
59	.32529	.34400	2.9070	.94561	1		59	.34175	.36364	2.7500	.93979	1
60	.32557	.34433	2.9042	.94552	0		60	.34202	.36397	2.7475	.93969	0
′	Cos	Cot	Tan	Sin	′		′	Cos	Cot	Tan	Sin	′

108° (288°) (251°) **71°**   **109° (289°)** (250°) **70°**

NATURAL TRIGONOMETRIC FUNCTIONS

20° (200°)   **(339°) 159°**

′	Sin	Tan	Cot	Cos	′
0	.34202	.36397	2.7475	.93969	60
1	.34229	.36430	2.7450	.93959	59
2	.34257	.36463	2.7425	.93949	58
3	.34284	.36496	2.7400	.93930	57
4	.34311	.36529	2.7376	.93929	56
5	.34339	.36562	2.7351	.93919	55
6	.34366	.36595	2.7326	.93909	54
7	.34393	.36628	2.7302	.93899	53
8	.34421	.36661	2.7277	.93889	52
9	.34448	.36694	2.7253	.93879	51
10	.34475	.36727	2.7228	.93869	50
11	.34503	.36760	2.7204	.93859	49
12	.34530	.36793	2.7170	.93849	48
13	.34557	.36826	2.7155	.93839	47
14	.34584	.36859	2.7130	.93829	46
15	.34612	.36892	2.7106	.93819	45
16	.34639	.36925	2.7082	.93809	44
17	.34666	.36958	2.7058	.93799	43
18	.34694	.36991	2.7034	.93789	42
19	.34721	.37024	2.7009	.93779	41
20	.34748	.37057	2.6985	.93769	40
21	.34775	.37090	2.6961	.93759	39
22	.34803	.37123	2.6937	.93748	38
23	.34830	.37157	2.6913	.93738	37
24	.34857	.37190	2.6889	.93728	36
25	.34884	.37223	2.6865	.93718	35
26	.34912	.37256	2.6841	.93708	34
27	.34939	.37289	2.6818	.93698	33
28	.34966	.37322	2.6794	.93688	32
29	.34993	.37355	2.6770	.93677	31
30	.35021	.37388	2.6746	.93667	30
31	.35048	.37422	2.6723	.93657	29
32	.35075	.37455	2.6699	.93647	28
33	.35102	.37488	2.6675	.93637	27
34	.35130	.37521	2.6652	.93626	26
35	.35157	.37554	2.6628	.93616	25
36	.35184	.37588	2.6605	.93606	24
37	.35211	.37621	2.6581	.93596	23
38	.35239	.37654	2.6558	.93585	22
39	.35266	.37687	2.6534	.93575	21
40	.35293	.37720	2.6511	.93565	20
41	.35320	.37754	2.6488	.93555	19
42	.35347	.37787	2.6464	.93544	18
43	.35375	.37820	2.6441	.93534	17
44	.35402	.37853	2.6418	.93524	16
45	.35429	.37887	2.6395	.93514	15
46	.35456	.37920	2.6371	.93503	14
47	.35484	.37953	2.6348	.93493	13
48	.35511	.37986	2.6325	.93483	12
49	.35538	.38020	2.6302	.93472	11
50	.35565	.38053	2.6279	.93462	10
51	.35592	.38086	2.6256	.93452	9
52	.35619	.38120	2.6233	.93441	8
53	.35647	.38153	2.6210	.93431	7
54	.35674	.38186	2.6187	.93420	6
55	.35701	.38220	2.6165	.93410	5
56	.35728	.38253	2.6142	.93400	4
57	.35755	.38286	2.6119	.93389	3
58	.35782	.38320	2.6096	.93379	2
59	.35810	.38353	2.6074	.93368	1
60	.35837	.38386	2.6051	.93358	0
′	Cos	Cot	Tan	Sin	′

110° (290°)   **(249°) 69°**

21° (201°)   **(338°) 158°**

′	Sin	Tan	Cot	Cos	′
0	.35837	.38386	2.6051	.93358	60
1	.35864	.38420	2.6028	.93348	59
2	.35891	.38453	2.6006	.93337	58
3	.35918	.38487	2.5983	.93327	57
4	.35945	.38520	2.5961	.93316	56
5	.35973	.38553	2.5938	.93306	55
6	.36000	.38587	2.5916	.93295	54
7	.36027	.38620	2.5893	.93285	53
8	.36054	.38654	2.5871	.93274	52
9	.36081	.38687	2.5848	.93264	51
10	.36108	.38721	2.5826	.93253	50
11	.36135	.38754	2.5804	.93243	49
12	.36162	.38787	2.5782	.93232	48
13	.36190	.38821	2.5759	.93222	47
14	.36217	.38854	2.5737	.93211	46
15	.36244	.38888	2.5715	.93201	45
16	.36271	.38921	2.5693	.93190	44
17	.36298	.38955	2.5671	.93180	43
18	.36325	.38988	2.5649	.93169	42
19	.36352	.39022	2.5627	.93159	41
20	.36379	.39055	2.5605	.93148	40
21	.36406	.39089	2.5583	.93137	39
22	.36434	.39122	2.5561	.93127	38
23	.36461	.39156	2.5539	.93116	37
24	.36488	.39190	2.5517	.93106	36
25	.36515	.39223	2.5495	.93095	35
26	.36542	.39257	2.5473	.93084	34
27	.36569	.39290	2.5452	.93074	33
28	.36596	.39324	2.5430	.93063	32
29	.36623	.39357	2.5408	.93052	31
30	.36650	.39391	2.5386	.93042	30
31	.36677	.39425	2.5365	.93031	29
32	.36704	.39458	2.5343	.93020	28
33	.36731	.39492	2.5322	.93010	27
34	.36758	.39526	2.5300	.92999	26
35	.36785	.39559	2.5279	.92988	25
36	.36812	.39593	2.5257	.92978	24
37	.36839	.39626	2.5236	.92967	23
38	.36867	.39660	2.5214	.92956	22
39	.36894	.39694	2.5193	.92945	21
40	.36921	.39727	2.5172	.92935	20
41	.36948	.39761	2.5150	.92924	19
42	.36975	.39795	2.5129	.92913	18
43	.37002	.39829	2.5108	.92902	17
44	.37029	.39862	2.5086	.92892	16
45	.37056	.39896	2.5065	.92881	15
46	.37083	.39930	2.5044	.92870	14
47	.37110	.39963	2.5023	.92859	13
48	.37137	.39997	2.5002	.92849	12
49	.37164	.40031	2.4981	.92838	11
50	.37191	.40065	2.4960	.92827	10
51	.37218	.40098	2.4939	.92816	9
52	.37245	.40132	2.4918	.92805	8
53	.37272	.40166	2.4897	.92794	7
54	.37299	.40200	2.4876	.92784	6
55	.37326	.40234	2.4855	.92773	5
56	.37353	.40267	2.4834	.92762	4
57	.37380	.40301	2.4813	.92751	3
58	.37407	.40335	2.4792	.92740	2
59	.37434	.40369	2.4772	.92729	1
60	.37461	.40403	2.4751	.92718	0
′	Cos	Cot	Tan	Sin	′

111° (291°)   **(248°) 68°**

NATURAL TRIGONOMETRIC FUNCTIONS

22° (202°)   **(337°) 157°**

'	Sin	Tan	Cot	Cos	'
0	.37461	.40403	2.4751	.92718	60
1	.37488	.40436	2.4730	.92707	59
2	.37515	.40470	2.4709	.92697	58
3	.37542	.40504	2.4689	.92686	57
4	.37569	.40538	2.4668	.92675	56
5	.37595	.40572	2.4648	.92664	55
6	.37622	.40606	2.4627	.92653	54
7	.37649	.40640	2.4606	.92642	53
8	.37676	.40674	2.4586	.92631	52
9	.37703	.40707	2.4566	.92620	51
10	.37730	.40741	2.4545	.92609	50
11	.37757	.40775	2.4525	.92598	49
12	.37784	.40809	2.4504	.92587	48
13	.37811	.40843	2.4484	.92576	47
14	.37838	.40877	2.4464	.92565	46
15	.37865	.40911	2.4443	.92554	45
16	.37892	.40945	2.4423	.92543	44
17	.37919	.40979	2.4403	.92532	43
18	.37946	.41013	2.4383	.92521	42
19	.37973	.41047	2.4362	.92510	41
20	.37999	.41081	2.4342	.92499	40
21	.38026	.41115	2.4322	.92488	39
22	.38053	.41149	2.4302	.92477	38
23	.38080	.41183	2.4282	.92466	37
24	.38107	.41217	2.4262	.92455	36
25	.38134	.41251	2.4242	.92444	35
26	.38161	.41285	2.4222	.92432	34
27	.38188	.41319	2.4202	.92421	33
28	.38215	.41353	2.4182	.92410	32
29	.38241	.41387	2.4162	.92399	31
30	.38268	.41421	2.4142	.92388	30
31	.38295	.41455	2.4122	.92377	29
32	.38322	.41490	2.4102	.92366	28
33	.38349	.41524	2.4083	.92355	27
34	.38376	.41558	2.4063	.92343	26
35	.38403	.41592	2.4043	.92332	25
36	.38430	.41626	2.4023	.92321	24
37	.38456	.41660	2.4004	.92310	23
38	.38483	.41694	2.3984	.92299	22
39	.38510	.41728	2.3964	.92287	21
40	.38537	.41763	2.3945	.92276	20
41	.38564	.41797	2.3925	.92265	19
42	.38591	.41831	2.3906	.92254	18
43	.38617	.41865	2.3886	.92243	17
44	.38644	.41899	2.3867	.92231	16
45	.38671	.41933	2.3847	.92220	15
46	.38698	.41968	2.3828	.92209	14
47	.38725	.42002	2.3808	.92198	13
48	.38752	.42036	2.3789	.92186	12
49	.38778	.42070	2.3770	.92175	11
50	.38805	.42105	2.3750	.92164	10
51	.38832	.42139	2.3731	.92152	9
52	.38859	.42173	2.3712	.92141	8
53	.38886	.42207	2.3693	.92130	7
54	.38912	.42242	2.3673	.92119	6
55	.38939	.42276	2.3654	.92107	5
56	.38966	.42310	2.3635	.92096	4
57	.38993	.42345	2.3616	.92085	3
58	.39020	.42379	2.3597	.92073	2
59	.39046	.42413	2.3578	.92062	1
60	.39073	.42447	2.3559	.92050	0
'	Cos	Cot	Tan	Sin	'

112° (292°)   **(247°) 67°**

23° (203°)   **(336°) 156°**

'	Sin	Tan	Cot	Cos	'
0	.39073	.42447	2.3559	.92050	60
1	.39100	.42482	2.3539	.92039	59
2	.39127	.42516	2.3520	.92028	58
3	.39153	.42551	2.3501	.92016	57
4	.39180	.42585	2.3483	.92005	56
5	.39207	.42619	2.3464	.91994	55
6	.39234	.42654	2.3445	.91982	54
7	.39260	.42688	2.3426	.91971	53
8	.39287	.42722	2.3407	.91959	52
9	.39314	.42757	2.3388	.91948	51
10	.39341	.42791	2.3369	.91936	50
11	.39367	.42826	2.3351	.91925	49
12	.39394	.42860	2.3332	.91914	48
13	.39421	.42894	2.3313	.91902	47
14	.39448	.42929	2.3294	.91891	46
15	.39474	.42963	2.3276	.91879	45
16	.39501	.42998	2.3257	.91868	44
17	.39528	.43032	2.3238	.91856	43
18	.39555	.43067	2.3220	.91845	42
19	.39581	.43101	2.3201	.91833	41
20	.39608	.43136	2.3183	.91822	40
21	.39635	.43170	2.3164	.91810	39
22	.39661	.43205	2.3146	.91799	38
23	.39688	.43239	2.3127	.91787	37
24	.39715	.43274	2.3109	.91775	36
25	.39741	.43308	2.3090	.91764	35
26	.39768	.43343	2.3072	.91752	34
27	.39795	.43378	2.3053	.91741	33
28	.39822	.43412	2.3035	.91729	32
29	.39848	.43447	2.3017	.91718	31
30	.39875	.43481	2.2998	.91706	30
31	.39902	.43516	2.2980	.91694	29
32	.39928	.43550	2.2962	.91683	28
33	.39955	.43585	2.2944	.91671	27
34	.39982	.43620	2.2925	.91660	26
35	.40008	.43654	2.2907	.91648	25
36	.40035	.43689	2.2889	.91636	24
37	.40062	.43724	2.2871	.91625	23
38	.40088	.43758	2.2853	.91613	22
39	.40115	.43793	2.2835	.91601	21
40	.40141	.43828	2.2817	.91590	20
41	.40168	.43862	2.2799	.91578	19
42	.40195	.43897	2.2781	.91566	18
43	.40221	.43932	2.2763	.91555	17
44	.40248	.43966	2.2745	.91543	16
45	.40275	.44001	2.2727	.91531	15
46	.40301	.44036	2.2709	.91519	14
47	.40328	.44071	2.2691	.91508	13
48	.40355	.44105	2.2673	.91496	12
49	.40381	.44140	2.2655	.91484	11
50	.40408	.44175	2.2637	.91472	10
51	.40434	.44210	2.2620	.91461	9
52	.40461	.44244	2.2602	.91449	8
53	.40488	.44279	2.2584	.91437	7
54	.40514	.44314	2.2566	.91425	6
55	.40541	.44349	2.2549	.91414	5
56	.40567	.44384	2.2531	.91402	4
57	.40594	.44418	2.2513	.91390	3
58	.40621	.44453	2.2496	.91378	2
59	.40647	.44488	2.2478	.91366	1
60	.40674	.44523	2.2460	.91355	0
'	Cos	Cot	Tan	Sin	'

113° (293°)   **(246°) 66°**

24° (204°)   (335°) **155°**   **25° (205°)**   (334°) **154°**

′	Sin	Tan	Cot	Cos	′		′	Sin	Tan	Cot	Cos	′
0	.40674	.44523	2.2460	.91355	60		0	.42262	.46631	2.1445	.90631	60
1	.40700	.44558	2.2443	.91343	59		1	.42288	.46666	2.1429	.90618	59
2	.40727	.44593	2.2425	.91331	58		2	.42315	.46702	2.1413	.90606	58
3	.40753	.44627	2.2408	.91319	57		3	.42341	.46737	2.1396	.90594	57
4	.40780	.44662	2.2390	.91307	56		4	.42367	.46772	2.1380	.90582	56
5	.40806	.44697	2.2373	.91295	55		5	.42394	.46808	2.1364	.90569	55
6	.40833	.44732	2.2355	.91283	54		6	.42420	.46843	2.1348	.90557	54
7	.40860	.44767	2.2338	.91272	53		7	.42446	.46879	2.1332	.90545	53
8	.40886	.44802	2.2320	.91260	52		8	.42473	.46914	2.1315	.90532	52
9	.40913	.44837	2.2303	.91248	51		9	.42499	.46950	2.1299	.90520	51
10	.40939	.44872	2.2286	.91236	50		10	.42525	.46985	2.1283	.90507	50
11	.40966	.44907	2.2268	.91224	49		11	.42552	.47021	2.1267	.90495	49
12	.40992	.44942	2.2251	.91212	48		12	.42578	.47056	2.1251	.90483	48
13	.41019	.44977	2.2234	.91200	47		13	.42604	.47092	2.1235	.90470	47
14	.41045	.45012	2.2216	.91188	46		14	.42631	.47128	2.1219	.90458	46
15	.41072	.45047	2.2199	.91176	45		15	.42657	.47163	2.1203	.90446	45
16	.41098	.45082	2.2182	.91164	44		16	.42683	.47199	2.1187	.90433	44
17	.41125	.45117	2.2165	.91152	43		17	.42709	.47234	2.1171	.90421	43
18	.41151	.45152	2.2148	.91140	42		18	.42736	.47270	2.1155	.90408	42
19	.41178	.45187	2.2130	.91128	41		19	.42762	.47305	2.1139	.90396	41
20	.41204	.45222	2.2113	.91116	40		20	.42788	.47341	2.1123	.90383	40
21	.41231	.45257	2.2096	.91104	39		21	.42815	.47377	2.1107	.90371	39
22	.41257	.45292	2.2079	.91092	38		22	.42841	.47412	2.1092	.90358	38
23	.41284	.45327	2.2062	.91080	37		23	.42867	.47448	2.1076	.90346	37
24	.41310	.45362	2.2045	.91068	36		24	.42894	.47483	2.1060	.90334	36
25	.41337	.45397	2.2028	.91056	35		25	.42920	.47519	2.1044	.90321	35
26	.41363	.45432	2.2011	.91044	34		26	.42946	.47555	2.1028	.90309	34
27	.41390	.45467	2.1994	.91032	33		27	.42972	.47590	2.1013	.90296	33
28	.41416	.45502	2.1977	.91020	32		28	.42999	.47626	2.0997	.90284	32
29	.41443	.45538	2.1960	.91008	31		29	.43025	.47662	2.0981	.90271	31
30	.41469	.45573	2.1943	.90996	30		30	.43051	.47698	2.0965	.90259	30
31	.41496	.45608	2.1926	.90984	29		31	.43077	.47733	2.0950	.90246	29
32	.41522	.45643	2.1909	.90972	28		32	.43104	.47769	2.0934	.90233	28
33	.41549	.45678	2.1892	.90960	27		33	.43130	.47805	2.0918	.90221	27
34	.41575	.45713	2.1876	.90948	26		34	.43156	.47840	2.0903	.90208	26
35	.41602	.45748	2.1859	.90936	25		35	.43182	.47876	2.0887	.90196	25
36	.41628	.45784	2.1842	.90924	24		36	.43209	.47912	2.0872	.90183	24
37	.41655	.45819	2.1825	.90911	23		37	.43235	.47948	2.0856	.90171	23
38	.41681	.45854	2.1808	.90899	22		38	.43261	.47984	2.0840	.90158	22
39	.41707	.45889	2.1792	.90887	21		39	.43287	.48019	2.0825	.90146	21
40	.41734	.45924	2.1775	.90875	20		40	.43313	.48055	2.0809	.90133	20
41	.41760	.45960	2.1758	.90863	19		41	.43340	.48091	2.0794	.90120	19
42	.41787	.45995	2.1742	.90851	18		42	.43366	.48127	2.0778	.90108	18
43	.41813	.46030	2.1725	.90839	17		43	.43392	.48163	2.0763	.90095	17
44	.41840	.46065	2.1708	.90826	16		44	.43418	.48198	2.0748	.90082	16
45	.41866	.46101	2.1692	.90814	15		45	.43445	.48234	2.0732	.90070	15
46	.41892	.46136	2.1675	.90802	14		46	.43471	.48270	2.0717	.90057	14
47	.41919	.46171	2.1659	.90790	13		47	.43497	.48306	2.0701	.90045	13
48	.41945	.46206	2.1642	.90778	12		48	.43523	.48342	2.0686	.90032	12
49	.41972	.46242	2.1625	.90766	11		49	.43549	.48378	2.0671	.90019	11
50	.41998	.46277	2.1609	.90753	10		50	.43575	.48414	2.0655	.90007	10
51	.42024	.46312	2.1592	.90741	9		51	.43602	.48450	2.0640	.89994	9
52	.42051	.46348	2.1576	.90729	8		52	.43628	.48486	2.0625	.89981	8
53	.42077	.46383	2.1560	.90717	7		53	.43654	.48521	2.0609	.89968	7
54	.42104	.46418	2.1543	.90704	6		54	.43680	.48557	2.0594	.89956	6
55	.42130	.46454	2.1527	.90692	5		55	.43706	.48593	2.0579	.89943	5
56	.42156	.46489	2.1510	.90680	4		56	.43733	.48629	2.0564	.89930	4
57	.42183	.46525	2.1494	.90668	3		57	.43759	.48665	2.0549	.89918	3
58	.42209	.46560	2.1478	.90655	2		58	.43785	.48701	2.0533	.89905	2
59	.42235	.46595	2.1461	.90643	1		59	.43811	.48737	2.0518	.89892	1
60	.42262	.46631	2.1445	.90631	0		60	.43837	.48773	2.0503	.89879	0
′	Cos	Cot	Tan	Sin	′		′	Cos	Cot	Tan	Sin	′

114° (294°)   (245°) **65°**   **115° (295°)**   (244°) **64°**

NATURAL TRIGONOMETRIC FUNCTIONS

26° (206°) (333°) 153°

′	Sin	Tan	Cot	Cos	′
0	.43837	.48773	2.0503	.89879	60
1	.43863	.48809	2.0488	.89867	59
2	.43889	.48845	2.0473	.89854	58
3	.43916	.48881	2.0458	.89841	57
4	.43942	.48917	2.0443	.89828	56
5	.43968	.48953	2.0428	.89816	55
6	.43994	.48989	2.0413	.89803	54
7	.44020	.49026	2.0398	.89790	53
8	.44046	.49062	2.0383	.89777	52
9	.44072	.49098	2.0368	.89764	51
10	.44098	.49134	2.0353	.89752	50
11	.44124	.49170	2.0338	.89739	49
12	.44151	.49206	2.0323	.89726	48
13	.44177	.49242	2.0308	.89713	47
14	.44203	.49278	2.0293	.89700	46
15	.44229	.49315	2.0278	.89687	45
16	.44255	.49351	2.0263	.89674	44
17	.44281	.49387	2.0248	.89662	43
18	.44307	.49423	2.0233	.89649	42
19	.44333	.49459	2.0219	.89636	41
20	.44359	.49495	2.0204	.89623	40
21	.44385	.49532	2.0189	.89610	39
22	.44411	.49568	2.0174	.89597	38
23	.44437	.49604	2.0160	.89584	37
24	.44464	.49640	2.0145	.89571	36
25	.44490	.49677	2.0130	.89558	35
26	.44516	.49713	2.0115	.89545	34
27	.44542	.49749	2.0101	.89532	33
28	.44568	.49786	2.0086	.89519	32
29	.44594	.49822	2.0072	.89506	31
30	.44620	.49858	2.0057	.89493	30
31	.44646	.49894	2.0042	.89480	29
32	.44672	.49931	2.0028	.89467	28
33	.44698	.49967	2.0013	.89454	27
34	.44724	.50004	1.9999	.89441	26
35	.44750	.50040	1.9984	.89428	25
36	.44776	.50076	1.9970	.89415	24
37	.44802	.50113	1.9955	.89402	23
38	.44828	.50149	1.9941	.89389	22
39	.44854	.50185	1.9926	.89376	21
40	.44880	.50222	1.9912	.89363	20
41	.44906	.50258	1.9897	.89350	19
42	.44932	.50295	1.9883	.89337	18
43	.44958	.50331	1.9868	.89324	17
44	.44984	.50368	1.9854	.89311	16
45	.45010	.50404	1.9840	.89298	15
46	.45036	.50441	1.9825	.89285	14
47	.45062	.50477	1.9811	.89272	13
48	.45088	.50514	1.9797	.89259	12
49	.45114	.50550	1.9782	.89245	11
50	.45140	.50587	1.9768	.89232	10
51	.45166	.50623	1.9754	.89219	9
52	.45192	.50660	1.9740	.89206	8
53	.45218	.50696	1.9725	.89193	7
54	.45243	.50733	1.9711	.89180	6
55	.45269	.50769	1.9697	.89167	5
56	.45295	.50806	1.9683	.89153	4
57	.45321	.50843	1.9669	.89140	3
58	.45347	.50879	1.9654	.89127	2
59	.45373	.50916	1.9640	.89114	1
60	.45399	.50953	1.9626	.89101	0
′	Cos	Cot	Tan	Sin	′

116° (296°) (243°) 63°

27° (207°) (332°) 152°

′	Sin	Tan	Cot	Cos	′
0	.45399	.50953	1.9626	.89101	60
1	.45425	.50989	1.9612	.89087	59
2	.45451	.51026	1.9598	.89074	58
3	.45477	.51063	1.9584	.89061	57
4	.45503	.51099	1.9570	.89048	56
5	.45529	.51136	1.9556	.89035	55
6	.45554	.51173	1.9542	.89021	54
7	.45580	.51209	1.9528	.89008	53
8	.45606	.51246	1.9514	.88995	52
9	.45632	.51283	1.9500	.88981	51
10	.45658	.51319	1.9486	.88968	50
11	.45684	.51356	1.9472	.88955	49
12	.45710	.51393	1.9458	.88942	48
13	.45736	.51430	1.9444	.88928	47
14	.45762	.51467	1.9430	.88915	46
15	.45787	.51503	1.9416	.88902	45
16	.45813	.51540	1.9402	.88888	44
17	.45839	.51577	1.9388	.88875	43
18	.45865	.51614	1.9375	.88862	42
19	.45891	.51651	1.9361	.88848	41
20	.45917	.51688	1.9347	.88835	40
21	.45942	.51724	1.9333	.88822	39
22	.45968	.51761	1.9319	.88808	38
23	.45994	.51798	1.9306	.88795	37
24	.46020	.51835	1.9292	.88782	36
25	.46046	.51872	1.9278	.88768	35
26	.46072	.51909	1.9265	.88755	34
27	.46097	.51946	1.9251	.88741	33
28	.46123	.51983	1.9237	.88728	32
29	.46149	.52020	1.9223	.88715	31
30	.46175	.52057	1.9210	.88701	30
31	.46201	.52094	1.9196	.88688	29
32	.46226	.52131	1.9183	.88674	28
33	.46252	.52168	1.9169	.88661	27
34	.46278	.52205	1.9155	.88647	26
35	.46304	.52242	1.9142	.88634	25
36	.46330	.52279	1.9128	.88620	24
37	.46355	.52316	1.9115	.88607	23
38	.46381	.52353	1.9101	.88593	22
39	.46407	.52390	1.9088	.88580	21
40	.46433	.52427	1.9074	.88566	20
41	.46458	.52464	1.9061	.88553	19
42	.46484	.52501	1.9047	.88539	18
43	.46510	.52538	1.9034	.88526	17
44	.46536	.52575	1.9020	.88512	16
45	.46561	.52613	1.9007	.88499	15
46	.46587	.52650	1.8993	.88485	14
47	.46613	.52687	1.8980	.88472	13
48	.46639	.52724	1.8967	.88458	12
49	.46664	.52761	1.8953	.88445	11
50	.46690	.52798	1.8940	.88431	10
51	.46716	.52836	1.8927	.88417	9
52	.46742	.52873	1.8913	.88404	8
53	.46767	.52910	1.8900	.88390	7
54	.46793	.52947	1.8887	.88377	6
55	.46819	.52985	1.8873	.88363	5
56	.46844	.53022	1.8860	.88349	4
57	.46870	.53059	1.8847	.88336	3
58	.46896	.53096	1.8834	.88322	2
59	.46921	.53134	1.8820	.88308	1
60	.46947	.53171	1.8807	.88295	0
′	Cos	Cot	Tan	Sin	′

117° (297°) (242°) 62°

28° (208°) (331°) 151°

′	Sin	Tan	Cot	Cos	′
0	.46947	.53171	1.8807	.88295	60
1	.46973	.53208	1.8794	.88281	59
2	.46999	.53246	1.8781	.88267	58
3	.47024	.53283	1.8768	.88254	57
4	.47050	.53320	1.8755	.88240	56
5	.47076	.53358	1.8741	.88226	55
6	.47101	.53395	1.8728	.88213	54
7	.47127	.53432	1.8715	.88199	53
8	.47153	.53470	1.8702	.88185	52
9	.47178	.53507	1.8689	.88172	51
10	.47204	.53545	1.8676	.88158	50
11	.47229	.53582	1.8663	.88144	49
12	.47255	.53620	1.8650	.88130	48
13	.47281	.53657	1.8637	.88117	47
14	.47306	.53694	1.8624	.88103	46
15	.47332	.53732	1.8611	.88089	45
16	.47358	.53769	1.8598	.88075	44
17	.47383	.53807	1.8585	.88062	43
18	.47409	.53844	1.8572	.88048	42
19	.47434	.53882	1.8559	.88034	41
20	.47460	.53920	1.8546	.88020	40
21	.47486	.53957	1.8533	.88006	39
22	.47511	.53995	1.8520	.87993	38
23	.47537	.54032	1.8507	.87979	37
24	.47562	.54070	1.8495	.87965	36
25	.47588	.54107	1.8482	.87951	35
26	.47614	.54145	1.8469	.87937	34
27	.47639	.54183	1.8456	.87923	33
28	.47665	.54220	1.8443	.87909	32
29	.47690	.54258	1.8430	.87896	31
30	.47716	.54296	1.8418	.87882	30
31	.47741	.54333	1.8405	.87868	29
32	.47767	.54371	1.8392	.87854	28
33	.47793	.54409	1.8379	.87840	27
34	.47818	.54446	1.8367	.87826	26
35	.47844	.54484	1.8354	.87812	25
36	.47869	.54522	1.8341	.87798	24
37	.47895	.54560	1.8329	.87784	23
38	.47920	.54597	1.8316	.87770	22
39	.47946	.54635	1.8303	.87756	21
40	.47971	.54673	1.8291	.87743	20
41	.47997	.54711	1.8278	.87729	19
42	.48022	.54748	1.8265	.87715	18
43	.48048	.54786	1.8253	.87701	17
44	.48073	.54824	1.8240	.87687	16
45	.48099	.54862	1.8228	.87673	15
46	.48124	.54900	1.8215	.87659	14
47	.48150	.54938	1.8202	.87645	13
48	.48175	.54975	1.8190	.87631	12
49	.48201	.55013	1.8177	.87617	11
50	.48226	.55051	1.8165	.87603	10
51	.48252	.55089	1.8152	.87589	9
52	.48277	.55127	1.8140	.87575	8
53	.48303	.55165	1.8127	.87561	7
54	.48328	.55203	1.8115	.87546	6
55	.48354	.55241	1.8103	.87532	5
56	.48379	.55279	1.8090	.87518	4
57	.48405	.55317	1.8078	.87504	3
58	.48430	.55355	1.8065	.87490	2
59	.48456	.55393	1.8053	.87476	1
60	.48481	.55431	1.8040	.87462	0
′	Cos	Cot	Tan	Sin	′

118° (298°) (241°) 61°

29° (209°) (330°) 150°

′	Sin	Tan	Cot	Cos	′
0	.48481	.55431	1.8040	.87462	60
1	.48506	.55469	1.8028	.87448	59
2	.48532	.55507	1.8016	.87434	58
3	.48557	.55545	1.8003	.87420	57
4	.48583	.55583	1.7991	.87406	56
5	.48608	.55621	1.7979	.87391	55
6	.48634	.55659	1.7966	.87377	54
7	.48659	.55697	1.7954	.87363	53
8	.48684	.55736	1.7942	.87349	52
9	.48710	.55774	1.7930	.87335	51
10	.48735	.55812	1.7917	.87321	50
11	.48761	.55850	1.7905	.87306	49
12	.48786	.55888	1.7893	.87292	48
13	.48811	.55926	1.7881	.87278	47
14	.48837	.55964	1.7868	.87264	46
15	.48862	.56003	1.7856	.87250	45
16	.48888	.56041	1.7844	.87235	44
17	.48913	.56079	1.7832	.87221	43
18	.48938	.56117	1.7820	.87207	42
19	.48964	.56156	1.7808	.87193	41
20	.48989	.56194	1.7796	.87178	40
21	.49014	.56232	1.7783	.87164	39
22	.49040	.56270	1.7771	.87150	38
23	.49065	.56309	1.7759	.87136	37
24	.49090	.56347	1.7747	.87121	36
25	.49116	.56385	1.7735	.87107	35
26	.49141	.56424	1.7723	.87093	34
27	.49166	.56462	1.7711	.87079	33
28	.49192	.56501	1.7699	.87064	32
29	.49217	.56539	1.7687	.87050	31
30	.49242	.56577	1.7675	.87036	30
31	.49268	.56616	1.7663	.87021	29
32	.49293	.56654	1.7651	.87007	28
33	.49318	.56693	1.7639	.86993	27
34	.49344	.56731	1.7627	.86978	26
35	.49369	.56769	1.7615	.86964	25
36	.49394	.56808	1.7603	.86949	24
37	.49419	.56846	1.7591	.86935	23
38	.49445	.56885	1.7579	.86921	22
39	.49470	.56923	1.7567	.86906	21
40	.49495	.56962	1.7556	.86892	20
41	.49521	.57000	1.7544	.86878	19
42	.49546	.57039	1.7532	.86863	18
43	.49571	.57078	1.7520	.86849	17
44	.49596	.57116	1.7508	.86834	16
45	.49622	.57155	1.7496	.86820	15
46	.49647	.57193	1.7485	.86805	14
47	.49672	.57232	1.7473	.86791	13
48	.49697	.57271	1.7461	.86777	12
49	.49723	.57309	1.7449	.86762	11
50	.49748	.57348	1.7437	.86748	10
51	.49773	.57386	1.7426	.86733	9
52	.49798	.57425	1.7414	.86719	8
53	.49824	.57464	1.7402	.86704	7
54	.49849	.57503	1.7391	.86690	6
55	.49874	.57541	1.7379	.86675	5
56	.49899	.57580	1.7367	.86661	4
57	.49924	.57619	1.7355	.86646	3
58	.49950	.57657	1.7344	.86632	2
59	.49975	.57696	1.7332	.86617	1
60	.50000	.57735	1.7321	.86603	0
′	Cos	Cot	Tan	Sin	′

119° (299°) (240°) 60°

NATURAL TRIGONOMETRIC FUNCTIONS

30° (210°)   **(329°) 149°**

'	Sin	Tan	Cot	Cos	'
0	.50000	.57735	1.7321	.86603	60
1	.50025	.57774	1.7309	.86588	59
2	.50050	.57813	1.7297	.86573	58
3	.50076	.57851	1.7286	.86559	57
4	.50101	.57890	1.7274	.86544	56
5	.50126	.57929	1.7262	.86530	55
6	.50151	.57968	1.7251	.86515	54
7	.50176	.58007	1.7239	.86501	53
8	.50201	.58046	1.7228	.86486	52
9	.50227	.58085	1.7216	.86471	51
10	.50252	.58124	1.7205	.86457	50
11	.50277	.58162	1.7193	.86442	49
12	.50302	.58201	1.7182	.86427	48
13	.50327	.58240	1.7170	.86413	47
14	.50352	.58279	1.7159	.86398	46
15	.50377	.58318	1.7147	.86384	45
16	.50403	.58357	1.7136	.86369	44
17	.50428	.58396	1.7124	.86354	43
18	.50453	.58435	1.7113	.86340	42
19	.50478	.58474	1.7102	.86325	41
20	.50503	.58513	1.7090	.86310	40
21	.50528	.58552	1.7079	.86295	39
22	.50553	.58591	1.7067	.86281	38
23	.50578	.58631	1.7056	.86266	37
24	.50603	.58670	1.7045	.86251	36
25	.50628	.58709	1.7033	.86237	35
26	.50654	.58748	1.7022	.86222	34
27	.50679	.58787	1.7011	.86207	33
28	.50704	.58826	1.6999	.86192	32
29	.50729	.58865	1.6988	.86178	31
30	.50754	.58905	1.6977	.86163	30
31	.50779	.58944	1.6965	.86148	29
32	.50804	.58983	1.6954	.86133	28
33	.50829	.59022	1.6943	.86119	27
34	.50854	.59061	1.6932	.86104	26
35	.50879	.59101	1.6920	.86089	25
36	.50904	.59140	1.6909	.86074	24
37	.50929	.59179	1.6898	.86059	23
38	.50954	.59218	1.6887	.86045	22
39	.50979	.59258	1.6875	.86030	21
40	.51004	.59297	1.6864	.86015	20
41	.51029	.59336	1.6853	.86000	19
42	.51054	.59376	1.6842	.85985	18
43	.51079	.59415	1.6831	.85970	17
44	.51104	.59454	1.6820	.85956	16
45	.51129	.59494	1.6808	.85941	15
46	.51154	.59533	1.6797	.85926	14
47	.51179	.59573	1.6786	.85911	13
48	.51204	.59612	1.6775	.85896	12
49	.51229	.59651	1.6764	.85881	11
50	.51254	.59691	1.6753	.85866	10
51	.51279	.59730	1.6742	.85851	9
52	.51304	.59770	1.6731	.85836	8
53	.51329	.59809	1.6720	.85821	7
54	.51354	.59849	1.6709	.85806	6
55	.51379	.59888	1.6698	.85792	5
56	.51404	.59928	1.6687	.85777	4
57	.51429	.59967	1.6676	.85762	3
58	.51454	.60007	1.6665	.85747	2
59	.51479	.60046	1.6654	.85732	1
60	.51504	.60086	1.6643	.85717	0
'	Cos	Cot	Tan	Sin	'

120° (300°)   **(239°) 59°**

31° (211°)   **(328°) 148°**

'	Sin	Tan	Cot	Cos	'
0	.51504	.60086	1.6643	.85717	60
1	.51529	.60126	1.6632	.85702	59
2	.51554	.60165	1.6621	.85687	58
3	.51579	.60205	1.6610	.85672	57
4	.51604	.60245	1.6599	.85657	56
5	.51628	.60284	1.6588	.85642	55
6	.51653	.60324	1.6577	.85627	54
7	.51678	.60364	1.6566	.85612	53
8	.51703	.60403	1.6555	.85597	52
9	.51728	.60443	1.6545	.85582	51
10	.51753	.60483	1.6534	.85567	50
11	.51778	.60522	1.6523	.85551	49
12	.51803	.60562	1.6512	.85536	48
13	.51828	.60602	1.6501	.85521	47
14	.51852	.60642	1.6490	.85506	46
15	.51877	.60681	1.6479	.85491	45
16	.51902	.60721	1.6469	.85476	44
17	.51927	.60761	1.6458	.85461	43
18	.51952	.60801	1.6447	.85446	42
19	.51977	.60841	1.6436	.85431	41
20	.52002	.60881	1.6426	.85416	40
21	.52026	.60921	1.6415	.85401	39
22	.52051	.60960	1.6404	.85385	38
23	.52076	.61000	1.6393	.85370	37
24	.52101	.61040	1.6383	.85355	36
25	.52126	.61080	1.6372	.85340	35
26	.52151	.61120	1.6361	.85325	34
27	.52175	.61160	1.6351	.85310	33
28	.52200	.61200	1.6340	.85294	32
29	.52225	.61240	1.6329	.85279	31
30	.52250	.61280	1.6319	.85264	30
31	.52275	.61320	1.6308	.85249	29
32	.52299	.61360	1.6297	.85234	28
33	.52324	.61400	1.6287	.85218	27
34	.52349	.61440	1.6276	.85203	26
35	.52374	.61480	1.6265	.85188	25
36	.52399	.61520	1.6255	.85173	24
37	.52423	.61561	1.6244	.85157	23
38	.52448	.61601	1.6234	.85142	22
39	.52473	.61641	1.6223	.85127	21
40	.52498	.61681	1.6212	.85112	20
41	.52522	.61721	1.6202	.85096	19
42	.52547	.61761	1.6191	.85081	18
43	.52572	.61801	1.6181	.85066	17
44	.52597	.61842	1.6170	.85051	16
45	.52621	.61882	1.6160	.85035	15
46	.52646	.61922	1.6149	.85020	14
47	.52671	.61962	1.6139	.85005	13
48	.52696	.62003	1.6128	.84989	12
49	.52720	.62043	1.6118	.84974	11
50	.52745	.62083	1.6107	.84959	10
51	.52770	.62124	1.6097	.84943	9
52	.52794	.62164	1.6087	.84928	8
53	.52819	.62204	1.6076	.84913	7
54	.52844	.62245	1.6066	.84897	6
55	.52869	.62285	1.6055	.84882	5
56	.52893	.62325	1.6045	.84866	4
57	.52918	.62366	1.6034	.84851	3
58	.52943	.62406	1.6024	.84836	2
59	.52967	.62446	1.6014	.84820	1
60	.52992	.62487	1.6003	.84805	0
'	Cos	Cot	Tan	Sin	'

121° (301°)   **(238°) 58°**

32° (212°)   **(327°) 147°**

′	Sin	Tan	Cot	Cos	′
0	.52992	.62487	1.6003	.84805	60
1	.53017	.62527	1.5993	.84789	59
2	.53041	.62568	1.5983	.84774	58
3	.53066	.62608	1.5972	.84759	57
4	.53091	.62649	1.5962	.84743	56
5	.53115	.62689	1.5952	.84728	55
6	.53140	.62730	1.5941	.84712	54
7	.53164	.62770	1.5931	.84697	53
8	.53189	.62811	1.5921	.84681	52
9	.53214	.62852	1.5911	.84666	51
10	.53238	.62892	1.5900	.84650	50
11	.53263	.62933	1.5890	.84635	49
12	.53288	.62973	1.5880	.84619	48
13	.53312	.63014	1.5869	.84604	47
14	.53337	.63055	1.5859	.84588	46
15	.53361	.63095	1.5849	.84573	45
16	.53386	.63136	1.5839	.84557	44
17	.53411	.63177	1.5829	.84542	43
18	.53435	.63217	1.5818	.84526	42
19	.53460	.63258	1.5808	.84511	41
20	.53484	.63299	1.5798	.84495	40
21	.53509	.63340	1.5788	.84480	39
22	.53534	.63380	1.5778	.84464	38
23	.53558	.63421	1.5768	.84448	37
24	.53583	.63462	1.5757	.84433	36
25	.53607	.63503	1.5747	.84417	35
26	.53632	.63544	1.5737	.84402	34
27	.53656	.63584	1.5727	.84386	33
28	.53681	.63625	1.5717	.84370	32
29	.53705	.63666	1.5707	.84355	31
30	.53730	.63707	1.5697	.84339	30
31	.53754	.63748	1.5687	.84324	29
32	.53779	.63789	1.5677	.84308	28
33	.53804	.63830	1.5667	.84292	27
34	.53828	.63871	1.5657	.84277	26
35	.53853	.63912	1.5647	.84261	25
36	.53877	.63953	1.5637	.84245	24
37	.53902	.63994	1.5627	.84230	23
38	.53926	.64035	1.5617	.84214	22
39	.53951	.64076	1.5607	.84198	21
40	.53975	.64117	1.5597	.84182	20
41	.54000	.64158	1.5587	.84167	19
42	.54024	.64199	1.5577	.84151	18
43	.54049	.64240	1.5567	.84135	17
44	.54073	.64281	1.5557	.84120	16
45	.54097	.64322	1.5547	.84104	15
46	.54122	.64363	1.5537	.84088	14
47	.54146	.64404	1.5527	.84072	13
48	.54171	.64446	1.5517	.84057	12
49	.54195	.64487	1.5507	.84041	11
50	.54220	.64528	1.5497	.84025	10
51	.54244	.64569	1.5487	.84009	9
52	.54269	.64610	1.5477	.83994	8
53	.54293	.64652	1.5468	.83978	7
54	.54317	.64693	1.5458	.83962	6
55	.54342	.64734	1.5448	.83946	5
56	.54366	.64775	1.5438	.83930	4
57	.54391	.64817	1.5428	.83915	3
58	.54415	.64858	1.5418	.83899	2
59	.54440	.64899	1.5408	.83883	1
60	.54464	.64941	1.5399	.83867	0
′	Cos	Cot	Tan	Sin	′

33° (213°)   **(326°) 146°**

′	Sin	Tan	Cot	Cos	′
0	.54464	.64941	1.5399	.83867	60
1	.54488	.64982	1.5389	.83851	59
2	.54513	.65024	1.5379	.83835	58
3	.54537	.65065	1.5369	.83819	57
4	.54561	.65106	1.5359	.83804	56
5	.54586	.65148	1.5350	.83788	55
6	.54610	.65189	1.5340	.83772	54
7	.54635	.65231	1.5330	.83756	53
8	.54659	.65272	1.5320	.83740	52
9	.54683	.65314	1.5311	.83724	51
10	.54708	.65355	1.5301	.83708	50
11	.54732	.65397	1.5291	.83692	49
12	.54756	.65438	1.5282	.83676	48
13	.54781	.65480	1.5272	.83660	47
14	.54805	.65521	1.5262	.83645	46
15	.54829	.65563	1.5253	.83629	45
16	.54854	.65604	1.5243	.83613	44
17	.54878	.65646	1.5233	.83597	43
18	.54902	.65688	1.5224	.83581	42
19	.54927	.65729	1.5214	.83565	41
20	.54951	.65771	1.5204	.83549	40
21	.54975	.65813	1.5195	.83533	39
22	.54999	.65854	1.5185	.83517	38
23	.55024	.65896	1.5175	.83501	37
24	.55048	.65938	1.5166	.83485	36
25	.55072	.65980	1.5156	.83469	35
26	.55097	.66021	1.5147	.83453	34
27	.55121	.66063	1.5137	.83437	33
28	.55145	.66105	1.5127	.83421	32
29	.55169	.66147	1.5118	.83405	31
30	.55194	.66189	1.5108	.83389	30
31	.55218	.66230	1.5099	.83373	29
32	.55242	.66272	1.5089	.83356	28
33	.55266	.66314	1.5080	.83340	27
34	.55291	.66356	1.5070	.83324	26
35	.55315	.66398	1.5061	.83308	25
36	.55339	.66440	1.5051	.83292	24
37	.55363	.66482	1.5042	.83276	23
38	.55388	.66524	1.5032	.83260	22
39	.55412	.66566	1.5023	.83244	21
40	.55436	.66608	1.5013	.83228	20
41	.55460	.66650	1.5004	.83212	19
42	.55484	.66692	1.4994	.83195	18
43	.55509	.66734	1.4985	.83179	17
44	.55533	.66776	1.4975	.83163	16
45	.55557	.66818	1.4966	.83147	15
46	.55581	.66860	1.4957	.83131	14
47	.55605	.66902	1.4947	.83115	13
48	.55630	.66944	1.4938	.83098	12
49	.55654	.66986	1.4928	.83082	11
50	.55678	.67028	1.4919	.83066	10
51	.55702	.67071	1.4910	.83050	9
52	.55726	.67113	1.4900	.83034	8
53	.55750	.67155	1.4891	.83017	7
54	.55775	.67197	1.4882	.83001	6
55	.55799	.67239	1.4872	.82985	5
56	.55823	.67282	1.4863	.82969	4
57	.55847	.67324	1.4854	.82953	3
58	.55871	.67366	1.4844	.82936	2
59	.55895	.67409	1.4835	.82920	1
60	.55919	.67451	1.4826	.82904	0
′	Cos	Cot	Tan	Sin	′

NATURAL TRIGONOMETRIC FUNCTIONS

34° (214°)   **(325°) 145°**

′	Sin	Tan	Cot	Cos	′
0	.55919	.67451	1.4826	.82904	60
1	.55943	.67493	1.4816	.82887	59
2	.55968	.67536	1.4807	.82871	58
3	.55992	.67578	1.4798	.82855	57
4	.56016	.67620	1.4788	.82839	56
5	.56040	.67663	1.4779	.82822	55
6	.56064	.67705	1.4770	.82806	54
7	.56088	.67748	1.4761	.82790	53
8	.56112	.67790	1.4751	.82773	52
9	.56136	.67832	1.4742	.82757	51
10	.56160	.67875	1.4733	.82741	50
11	.56184	.67917	1.4724	.82724	49
12	.56208	.67960	1.4715	.82708	48
13	.56232	.68002	1.4705	.82692	47
14	.56256	.68045	1.4696	.82675	46
15	.56280	.68088	1.4687	.82659	45
16	.56305	.68130	1.4678	.82643	44
17	.56329	.68173	1.4669	.82626	43
18	.56353	.68215	1.4659	.82610	42
19	.56377	.68258	1.4650	.82593	41
20	.56401	.68301	1.4641	.82577	40
21	.56425	.68343	1.4632	.82561	39
22	.56449	.68386	1.4623	.82544	38
23	.56473	.68429	1.4614	.82528	37
24	.56497	.68471	1.4605	.82511	36
25	.56521	.68514	1.4596	.82495	35
26	.56545	.68557	1.4586	.82478	34
27	.56569	.68600	1.4577	.82462	33
28	.56593	.68642	1.4568	.82446	32
29	.56617	.68685	1.4559	.82429	31
30	.56641	.68728	1.4550	.82413	30
31	.56665	.68771	1.4541	.82396	29
32	.56689	.68814	1.4532	.82380	28
33	.56713	.68857	1.4523	.82363	27
34	.56736	.68900	1.4514	.82347	26
35	.56760	.68942	1.4505	.82330	25
36	.56784	.68985	1.4496	.82314	24
37	.56808	.69028	1.4487	.82297	23
38	.56832	.69071	1.4478	.82281	22
39	.56856	.69114	1.4469	.82264	21
40	.56880	.69157	1.4460	.82248	20
41	.56904	.69200	1.4451	.82231	19
42	.56928	.69243	1.4442	.82214	18
43	.56952	.69286	1.4433	.82198	17
44	.56976	.69329	1.4424	.82181	16
45	.57000	.69372	1.4415	.82165	15
46	.57024	.69416	1.4406	.82148	14
47	.57047	.69459	1.4397	.82132	13
48	.57071	.69502	1.4388	.82115	12
49	.57095	.69545	1.4379	.82098	11
50	.57119	.69588	1.4370	.82082	10
51	.57143	.69631	1.4361	.82065	9
52	.57167	.69675	1.4352	.82048	8
53	.57191	.69718	1.4344	.82032	7
54	.57215	.69761	1.4335	.82015	6
55	.57238	.69804	1.4326	.81999	5
56	.57262	.69847	1.4317	.81982	4
57	.57286	.69891	1.4308	.81965	3
58	.57310	.69934	1.4299	.81949	2
59	.57334	.69977	1.4290	.81932	1
60	.57358	.70021	1.4281	.81915	0
′	Cos	Cot	Tan	Sin	′

124° (304°)   **(235°) 55°**

35° (215°)   **(324°) 144°**

′	Sin	Tan	Cot	Cos	′
0	.57358	.70021	1.4281	.81915	60
1	.57381	.70064	1.4273	.81899	59
2	.57405	.70107	1.4264	.81882	58
3	.57429	.70151	1.4255	.81865	57
4	.57453	.70194	1.4246	.81848	56
5	.57477	.70238	1.4237	.81832	55
6	.57501	.70281	1.4229	.81815	54
7	.57524	.70325	1.4220	.81798	53
8	.57548	.70368	1.4211	.81782	52
9	.57572	.70412	1.4202	.81765	51
10	.57596	.70455	1.4193	.81748	50
11	.57619	.70499	1.4185	.81731	49
12	.57643	.70542	1.4176	.81714	48
13	.57667	.70586	1.4167	.81698	47
14	.57691	.70629	1.4158	.81681	46
15	.57715	.70673	1.4150	.81664	45
16	.57738	.70717	1.4141	.81647	44
17	.57762	.70760	1.4132	.81631	43
18	.57786	.70804	1.4124	.81614	42
19	.57810	.70848	1.4115	.81597	41
20	.57833	.70891	1.4106	.81580	40
21	.57857	.70935	1.4097	.81563	39
22	.57881	.70979	1.4089	.81546	38
23	.57904	.71023	1.4080	.81530	37
24	.57928	.71066	1.4071	.81513	36
25	.57952	.71110	1.4063	.81496	35
26	.57976	.71154	1.4054	.81479	34
27	.57999	.71198	1.4045	.81462	33
28	.58023	.71242	1.4037	.81445	32
29	.58047	.71285	1.4028	.81428	31
30	.58070	.71329	1.4019	.81412	30
31	.58094	.71373	1.4011	.81395	29
32	.58118	.71417	1.4002	.81378	28
33	.58141	.71461	1.3994	.81361	27
34	.58165	.71505	1.3985	.81344	26
35	.58189	.71549	1.3976	.81327	25
36	.58212	.71593	1.3968	.81310	24
37	.58236	.71637	1.3959	.81293	23
38	.58260	.71681	1.3951	.81276	22
39	.58283	.71725	1.3942	.81259	21
40	.58307	.71769	1.3934	.81242	20
41	.58330	.71813	1.3925	.81225	19
42	.58354	.71857	1.3916	.81208	18
43	.58378	.71901	1.3908	.81191	17
44	.58401	.71946	1.3899	.81174	16
45	.58425	.71990	1.3891	.81157	15
46	.58449	.72034	1.3882	.81140	14
47	.58472	.72078	1.3874	.81123	13
48	.58496	.72122	1.3865	.81106	12
49	.58519	.72167	1.3857	.81089	11
50	.58543	.72211	1.3848	.81072	10
51	.58567	.72255	1.3840	.81055	9
52	.58590	.72299	1.3831	.81038	8
53	.58614	.72344	1.3823	.81021	7
54	.58637	.72388	1.3814	.81004	6
55	.58661	.72432	1.3806	.80987	5
56	.58684	.72477	1.3798	.80970	4
57	.58708	.72521	1.3789	.80953	3
58	.58731	.72565	1.3781	.80936	2
59	.58755	.72610	1.3772	.80919	1
60	.58779	.72654	1.3764	.80902	0
′	Cos	Cot	Tan	Sin	′

125° (305°)   **(234°) 54°**

36° (216°) (323°) **143°**

′	Sin	Tan	Cot	Cos	′
0	.58779	.72654	1.3764	.80902	60
1	.58802	.72699	1.3755	.80885	59
2	.58826	.72743	1.3747	.80867	58
3	.58849	.72788	1.3739	.80850	57
4	.58873	.72832	1.3730	.80833	56
5	.58896	.72877	1.3722	.80816	55
6	.58920	.72921	1.3713	.80799	54
7	.58943	.72966	1.3705	.80782	53
8	.58967	.73010	1.3697	.80765	52
9	.58990	.73055	1.3688	.80748	51
10	.59014	.73100	1.3680	.80730	50
11	.59037	.73144	1.3672	.80713	49
12	.59061	.73189	1.3663	.80696	48
13	.59084	.73234	1.3655	.80670	47
14	.59108	.73278	1.3647	.80662	46
15	.59131	.73323	1.3638	.80644	45
16	.59154	.73368	1.3630	.80627	44
17	.59178	.73413	1.3622	.80610	43
18	.59201	.73457	1.3613	.80593	42
19	.59225	.73502	1.3605	.80576	41
20	.59248	.73547	1.3597	.80558	40
21	.59272	.73592	1.3588	.80541	39
22	.59295	.73637	1.3580	.80524	38
23	.59318	.73681	1.3572	.80507	37
24	.59342	.73726	1.3564	.80489	36
25	.59365	.73771	1.3555	.80472	35
26	.59389	.73816	1.3547	.80455	34
27	.59412	.73861	1.3539	.80438	33
28	.59436	.73906	1.3531	.80420	32
29	.59459	.73951	1.3522	.80403	31
30	.59482	.73996	1.3514	.80386	30
31	.59506	.74041	1.3506	.80368	29
32	.59529	.74086	1.3498	.80351	28
33	.59552	.74131	1.3490	.80334	27
34	.59576	.74176	1.3481	.80316	26
35	.59599	.74221	1.3473	.80299	25
36	.59622	.74267	1.3465	.80282	24
37	.59646	.74312	1.3457	.80264	23
38	.59669	.74357	1.3449	.80247	22
39	.59693	.74402	1.3440	.80230	21
40	.59716	.74447	1.3432	.80212	20
41	.59739	.74492	1.3424	.80195	19
42	.59763	.74538	1.3416	.80178	18
43	.59786	.74583	1.3408	.80160	17
44	.59809	.74628	1.3400	.80143	16
45	.59832	.74674	1.3392	.80125	15
46	.59856	.74719	1.3384	.80108	14
47	.59879	.74764	1.3375	.80091	13
48	.59902	.74810	1.3367	.80073	12
49	.59926	.74855	1.3359	.80056	11
50	.59949	.74900	1.3351	.80038	10
51	.59972	.74946	1.3343	.80021	9
52	.59995	.74991	1.3335	.80003	8
53	.60019	.75037	1.3327	.79986	7
54	.60042	.75082	1.3319	.79968	6
55	.60065	.75128	1.3311	.79951	5
56	.60089	.75173	1.3303	.79934	4
57	.60112	.75219	1.3295	.79916	3
58	.60135	.75264	1.3287	.79899	2
59	.60158	.75310	1.3278	.79881	1
60	.60182	.75355	1.3270	.79864	0
′	Cos	Cot	Tan	Sin	′

126° (306°) (233°) **53°**

37° (217°) (322°) **142°**

′	Sin	Tan	Cot	Cos	′
0	.60182	.75355	1.3270	.79864	60
1	.60205	.75401	1.3262	.79846	59
2	.60228	.75447	1.3254	.79829	58
3	.60251	.75492	1.3246	.79811	57
4	.60274	.75538	1.3238	.79793	56
5	.60298	.75584	1.3230	.79776	55
6	.60321	.75629	1.3222	.79758	54
7	.60344	.75675	1.3214	.79741	53
8	.60367	.75721	1.3206	.79723	52
9	.60390	.75767	1.3198	.79706	51
10	.60414	.75812	1.3190	.79688	50
11	.60437	.75858	1.3182	.79671	49
12	.60460	.75904	1.3175	.79653	48
13	.60483	.75950	1.3167	.79635	47
14	.60506	.75996	1.3159	.79618	46
15	.60529	.76042	1.3151	.79600	45
16	.60553	.76088	1.3143	.79583	44
17	.60576	.76134	1.3135	.79565	43
18	.60599	.76180	1.3127	.79547	42
19	.60622	.76226	1.3119	.79530	41
20	.60645	.76272	1.3111	.79512	40
21	.60668	.76318	1.3103	.79494	39
22	.60691	.76364	1.3095	.79477	38
23	.60714	.76410	1.3087	.79459	37
24	.60738	.76456	1.3079	.79441	36
25	.60761	.76502	1.3072	.79424	35
26	.60784	.76548	1.3064	.79406	34
27	.60807	.76594	1.3056	.79388	33
28	.60830	.76640	1.3048	.79371	32
29	.60853	.76686	1.3040	.79353	31
30	.60876	.76733	1.3032	.79335	30
31	.60899	.76779	1.3024	.79318	29
32	.60922	.76825	1.3017	.79300	28
33	.60945	.76871	1.3009	.79282	27
34	.60968	.76918	1.3001	.79264	26
35	.60991	.76964	1.2993	.79247	25
36	.61015	.77010	1.2985	.79229	24
37	.61038	.77057	1.2977	.79211	23
38	.61061	.77103	1.2970	.79193	22
39	.61084	.77149	1.2962	.79176	21
40	.61107	.77196	1.2954	.79158	20
41	.61130	.77242	1.2946	.79140	19
42	.61153	.77289	1.2938	.79122	18
43	.61176	.77335	1.2931	.79105	17
44	.61199	.77382	1.2923	.79087	16
45	.61222	.77428	1.2915	.79069	15
46	.61245	.77475	1.2907	.79051	14
47	.61268	.77521	1.2900	.79033	13
48	.61291	.77568	1.2892	.79016	12
49	.61314	.77615	1.2884	.78998	11
50	.61337	.77661	1.2876	.78980	10
51	.61360	.77708	1.2869	.78962	9
52	.61383	.77754	1.2861	.78944	8
53	.61406	.77801	1.2853	.78926	7
54	.61429	.77848	1.2846	.78908	6
55	.61451	.77895	1.2838	.78891	5
56	.61474	.77941	1.2830	.78873	4
57	.61497	.77988	1.2822	.78855	3
58	.61520	.78035	1.2815	.78837	2
59	.61543	.78082	1.2807	.78819	1
60	.61566	.78129	1.2799	.78801	0
′	Cos	Cot	Tan	Sin	′

127° (307°) (232°) **52°**

NATURAL TRIGONOMETRIC FUNCTIONS

38° (218°) (321°) **141°**

′	Sin	Tan	Cot	Cos	′
0	.61566	.78129	1.2799	.78801	60
1	.61589	.78175	1.2792	.78783	59
2	.61612	.78222	1.2784	.78765	58
3	.61635	.78269	1.2776	.78747	57
4	.61658	.78316	1.2769	.78729	56
5	.61681	.78363	1.2761	.78711	55
6	.61704	.78410	1.2753	.78694	54
7	.61726	.78457	1.2746	.78676	53
8	.61749	.78504	1.2738	.78658	52
9	.61772	.78551	1.2731	.78640	51
10	.61795	.78598	1.2723	.78622	50
11	.61818	.78645	1.2715	.78604	49
12	.61841	.78692	1.2708	.78586	48
13	.61864	.78739	1.2700	.78568	47
14	.61887	.78786	1.2693	.78550	46
15	.61909	.78834	1.2685	.78532	45
16	.61932	.78881	1.2677	.78514	44
17	.61955	.78928	1.2670	.78496	43
18	.61978	.78975	1.2662	.78478	42
19	.62001	.79022	1.2655	.78460	41
20	.62024	.79070	1.2647	.78442	40
21	.62046	.79117	1.2640	.78424	39
22	.62069	.79164	1.2632	.78405	38
23	.62092	.79212	1.2624	.78387	37
24	.62115	.79259	1.2617	.78369	36
25	.62138	.79306	1.2609	.78351	35
26	.62160	.79354	1.2602	.78333	34
27	.62183	.79401	1.2594	.78315	33
28	.62206	.79449	1.2587	.78297	32
29	.62229	.79496	1.2579	.78279	31
30	.62251	.79544	1.2572	.78261	30
31	.62274	.79591	1.2564	.78243	29
32	.62297	.79639	1.2557	.78225	28
33	.62320	.79686	1.2549	.78206	27
34	.62342	.79734	1.2542	.78188	26
35	.62365	.79781	1.2534	.78170	25
36	.62388	.79829	1.2527	.78152	24
37	.62411	.79877	1.2519	.78134	23
38	.62433	.79924	1.2512	.78116	22
39	.62456	.79972	1.2504	.78098	21
40	.62479	.80020	1.2497	.78079	20
41	.62502	.80067	1.2489	.78061	19
42	.62524	.80115	1.2482	.78043	18
43	.62547	.80163	1.2475	.78025	17
44	.62570	.80211	1.2467	.78007	16
45	.62592	.80258	1.2460	.77988	15
46	.62615	.80306	1.2452	.77970	14
47	.62638	.80354	1.2445	.77952	13
48	.62660	.80402	1.2437	.77934	12
49	.62683	.80450	1.2430	.77916	11
50	.62706	.80498	1.2423	.77897	10
51	.62728	.80546	1.2415	.77879	9
52	.62751	.80594	1.2408	.77861	8
53	.62774	.80642	1.2401	.77843	7
54	.62796	.80690	1.2393	.77824	6
55	.62819	.80738	1.2386	.77806	5
56	.62842	.80786	1.2378	.77788	4
57	.62864	.80834	1.2371	.77769	3
58	.62887	.80882	1.2364	.77751	2
59	.62909	.80930	1.2356	.77733	1
60	.62932	.80978	1.2349	.77715	0
′	Cos	Cot	Tan	Sin	′

128° (308°) (231°) **51°**

39° (219°) (320°) **14[0]**

′	Sin	Tan	Cot	Cos	′
0	.62932	.80978	1.2349	.77715	60
1	.62955	.81027	1.2342	.77696	59
2	.62977	.81075	1.2334	.77678	58
3	.63000	.81123	1.2327	.77660	57
4	.63022	.81171	1.2320	.77641	56
5	.63045	.81220	1.2312	.77623	55
6	.63068	.81268	1.2305	.77605	54
7	.63090	.81316	1.2298	.77586	53
8	.63113	.81364	1.2290	.77568	52
9	.63135	.81413	1.2283	.77550	51
10	.63158	.81461	1.2276	.77531	50
11	.63180	.81510	1.2268	.77513	49
12	.63203	.81558	1.2261	.77494	48
13	.63225	.81606	1.2254	.77476	47
14	.63248	.81655	1.2247	.77458	46
15	.63271	.81703	1.2239	.77439	45
16	.63293	.81752	1.2232	.77421	44
17	.63316	.81800	1.2225	.77402	43
18	.63338	.81849	1.2218	.77384	42
19	.63361	.81898	1.2210	.77366	41
20	.63383	.81946	1.2203	.77347	40
21	.63406	.81995	1.2196	.77329	39
22	.63428	.82044	1.2189	.77310	38
23	.63451	.82092	1.2181	.77292	37
24	.63473	.82141	1.2174	.77273	36
25	.63496	.82190	1.2167	.77255	35
26	.63518	.82238	1.2160	.77236	34
27	.63540	.82287	1.2153	.77218	33
28	.63563	.82336	1.2145	.77199	32
29	.63585	.82385	1.2138	.77181	31
30	.63608	.82434	1.2131	.77162	30
31	.63630	.82483	1.2124	.77144	29
32	.63653	.82531	1.2117	.77125	28
33	.63675	.82580	1.2109	.77107	27
34	.63698	.82629	1.2102	.77088	26
35	.63720	.82678	1.2095	.77070	25
36	.63742	.82727	1.2088	.77051	24
37	.63765	.82776	1.2081	.77033	23
38	.63787	.82825	1.2074	.77014	22
39	.63810	.82874	1.2066	.76996	21
40	.63832	.82923	1.2059	.76977	20
41	.63854	.82972	1.2052	.76959	19
42	.63877	.83022	1.2045	.76940	18
43	.63899	.83071	1.2038	.76921	17
44	.63922	.83120	1.2031	.76903	16
45	.63944	.83169	1.2024	.76884	15
46	.63966	.83218	1.2017	.76866	14
47	.63989	.83268	1.2009	.76847	13
48	.64011	.83317	1.2002	.76828	12
49	.64033	.83366	1.1995	.76810	11
50	.64056	.83415	1.1988	.76791	10
51	.64078	.83465	1.1981	.76772	9
52	.64100	.83514	1.1974	.76754	8
53	.64123	.83564	1.1967	.76735	7
54	.64145	.83613	1.1960	.76717	6
55	.64167	.83662	1.1953	.76698	5
56	.64190	.83712	1.1946	.76679	4
57	.64212	.83761	1.1939	.76661	3
58	.64234	.83811	1.1932	.76642	2
59	.64256	.83860	1.1925	.76623	1
60	.64279	.83910	1.1918	.76604	0
′	Cos	Cot	Tan	Sin	′

129° (309°) (230°) **5[0°]**

40° (220°)   **(319°) 139°**

′	Sin	Tan	Cot	Cos	′
0	.64279	.83910	1.1918	.76604	60
1	.64301	.83960	1.1910	.76586	59
2	.64323	.84009	1.1903	.76567	58
3	.64346	.84059	1.1896	.76548	57
4	.64368	.84108	1.1889	.76530	56
5	.64390	.84158	1.1882	.76511	55
6	.64412	.84208	1.1875	.76492	54
7	.64435	.84258	1.1868	.76473	53
8	.64457	.84307	1.1861	.76455	52
9	.64479	.84357	1.1854	.76436	51
10	.64501	.84407	1.1847	.76417	50
11	.64524	.84457	1.1840	.76398	49
12	.64546	.84507	1.1833	.76380	48
13	.64568	.84556	1.1826	.76361	47
14	.64590	.84606	1.1819	.76342	46
15	.64612	.84656	1.1812	.76323	45
16	.64635	.84706	1.1806	.76304	44
17	.64657	.84756	1.1799	.76286	43
18	.64679	.84806	1.1792	.76267	42
19	.64701	.84856	1.1785	.76248	41
20	.64723	.84906	1.1778	.76229	40
21	.64746	.84956	1.1771	.76210	39
22	.64768	.85006	1.1764	.76192	38
23	.64790	.85057	1.1757	.76173	37
24	.64812	.85107	1.1750	.76154	36
25	.64834	.85157	1.1743	.76135	35
26	.64856	.85207	1.1736	.76116	34
27	.64878	.85257	1.1729	.76097	33
28	.64901	.85308	1.1722	.76078	32
29	.64923	.85358	1.1715	.76059	31
30	.64945	.85408	1.1708	.76041	30
31	.64967	.85458	1.1702	.76022	29
32	.64989	.85509	1.1695	.76003	28
33	.65011	.85559	1.1688	.75984	27
34	.65033	.85609	1.1681	.75965	26
35	.65055	.85660	1.1674	.75946	25
36	.65077	.85710	1.1667	.75927	24
37	.65100	.85761	1.1660	.75908	23
38	.65122	.85811	1.1653	.75889	22
39	.65144	.85862	1.1647	.75870	21
40	.65166	.85912	1.1640	.75851	20
41	.65188	.85963	1.1633	.75832	19
42	.65210	.86014	1.1626	.75813	18
43	.65232	.86064	1.1619	.75794	17
44	.65254	.86115	1.1612	.75775	16
45	.65276	.86166	1.1606	.75756	15
46	.65298	.86216	1.1599	.75738	14
47	.65320	.86267	1.1592	.75719	13
48	.65342	.86318	1.1585	.75700	12
49	.65364	.86368	1.1578	.75680	11
50	.65386	.86419	1.1571	.75661	10
51	.65408	.86470	1.1565	.75642	9
52	.65430	.86521	1.1558	.75623	8
53	.65452	.86572	1.1551	.75604	7
54	.65474	.86623	1.1544	.75585	6
55	.65496	.86674	1.1538	.75566	5
56	.65518	.86725	1.1531	.75547	4
57	.65540	.86776	1.1524	.75528	3
58	.65562	.86827	1.1517	.75509	2
59	.65584	.86878	1.1510	.75490	1
60	.65606	.86929	1.1504	.75471	0
′	Cos	Cot	Tan	Sin	′

130° (310°)   **(229°) 49°**

41° (221°)   **(318°) 138°**

′	Sin	Tan	Cot	Cos	′
0	.65606	.86929	1.1504	.75471	60
1	.65628	.86980	1.1497	.75452	59
2	.65650	.87031	1.1490	.75433	58
3	.65672	.87082	1.1483	.75414	57
4	.65694	.87133	1.1477	.75395	56
5	.65716	.87184	1.1470	.75375	55
6	.65738	.87236	1.1463	.75356	54
7	.65759	.87287	1.1456	.75337	53
8	.65781	.87338	1.1450	.75318	52
9	.65803	.87389	1.1443	.75299	51
10	.65825	.87441	1.1436	.75280	50
11	.65847	.87492	1.1430	.75261	49
12	.65869	.87543	1.1423	.75241	48
13	.65891	.87595	1.1416	.75222	47
14	.65913	.87646	1.1410	.75203	46
15	.65935	.87698	1.1403	.75184	45
16	.65956	.87749	1.1396	.75165	44
17	.65978	.87801	1.1389	.75146	43
18	.66000	.87852	1.1383	.75126	42
19	.66022	.87904	1.1376	.75107	41
20	.66044	.87955	1.1369	.75088	40
21	.66066	.88007	1.1363	.75069	39
22	.66088	.88059	1.1356	.75050	38
23	.66109	.88110	1.1349	.75030	37
24	.66131	.88162	1.1343	.75011	36
25	.66153	.88214	1.1336	.74992	35
26	.66175	.88265	1.1329	.74973	34
27	.66197	.88317	1.1323	.74953	33
28	.66218	.88369	1.1316	.74934	32
29	.66240	.88421	1.1310	.74915	31
30	.66262	.88473	1.1303	.74896	30
31	.66284	.88524	1.1296	.74876	29
32	.66306	.88576	1.1290	.74857	28
33	.66327	.88628	1.1283	.74838	27
34	.66349	.88680	1.1276	.74818	26
35	.66371	.88732	1.1270	.74799	25
36	.66393	.88784	1.1263	.74780	24
37	.66414	.88836	1.1257	.74760	23
38	.66436	.88888	1.1250	.74741	22
39	.66458	.88940	1.1243	.74722	21
40	.66480	.88992	1.1237	.74703	20
41	.66501	.89045	1.1230	.74683	19
42	.66523	.89097	1.1224	.74664	18
43	.66545	.89149	1.1217	.74644	17
44	.66566	.89201	1.1211	.74625	16
45	.66588	.89253	1.1204	.74606	15
46	.66610	.89306	1.1197	.74586	14
47	.66632	.89358	1.1191	.74567	13
48	.66653	.89410	1.1184	.74548	12
49	.66675	.89463	1.1178	.74528	11
50	.66697	.89515	1.1171	.74509	10
51	.66718	.89567	1.1165	.74489	9
52	.66740	.89620	1.1158	.74470	8
53	.66762	.89672	1.1152	.74451	7
54	.66783	.89725	1.1145	.74431	6
55	.66805	.89777	1.1139	.74412	5
56	.66827	.89830	1.1132	.74392	4
57	.66848	.89883	1.1126	.74373	3
58	.66870	.89935	1.1119	.74353	2
59	.66891	.89988	1.1113	.74334	1
60	.66913	.90040	1.1106	.74314	0
′	Cos	Cot	Tan	Sin	′

131° (311°)   **(228°) 48°**

42° (222°)   **(317°) 137°**

′	Sin	Tan	Cot	Cos	′
0	.66913	.90040	1.1106	.74314	60
1	.66935	.90093	1.1100	.74295	59
2	.66956	.90146	1.1093	.74276	58
3	.66978	.90199	1.1087	.74256	57
4	.66999	.90251	1.1080	.74237	56
5	.67021	.90304	1.1074	.74217	55
6	.67043	.90357	1.1067	.74198	54
7	.67064	.90410	1.1061	.74178	53
8	.67086	.90463	1.1054	.74159	52
9	.67107	.90516	1.1048	.74139	51
10	.67129	.90569	1.1041	.74120	50
11	.67151	.90621	1.1035	.74100	49
12	.67172	.90674	1.1028	.74080	48
13	.67194	.90727	1.1022	.74061	47
14	.67215	.90781	1.1016	.74041	46
15	.67237	.90834	1.1009	.74022	45
16	.67258	.90887	1.1003	.74002	44
17	.67280	.90940	1.0996	.73983	43
18	.67301	.90993	1.0990	.73963	42
19	.67323	.91046	1.0983	.73944	41
20	.67344	.91099	1.0977	.73924	40
21	.67366	.91153	1.0971	.73904	39
22	.67387	.91206	1.0964	.73885	38
23	.67409	.91259	1.0958	.73865	37
24	.67430	.91313	1.0951	.73846	36
25	.67452	.91366	1.0945	.73826	35
26	.67473	.91419	1.0939	.73806	34
27	.67495	.91473	1.0932	.73787	33
28	.67516	.91526	1.0926	.73767	32
29	.67538	.91580	1.0919	.73747	31
30	.67559	.91633	1.0913	.73728	30
31	.67580	.91687	1.0907	.73708	29
32	.67602	.91740	1.0900	.73688	28
33	.67623	.91794	1.0894	.73669	27
34	.67645	.91847	1.0888	.73649	26
35	.67666	.91901	1.0881	.73629	25
36	.67688	.91955	1.0875	.73610	24
37	.67709	.92008	1.0869	.73590	23
38	.67730	.92062	1.0862	.73570	22
39	.67752	.92116	1.0856	.73551	21
40	.67773	.92170	1.0850	.73531	20
41	.67795	.92224	1.0843	.73511	19
42	.67816	.92277	1.0837	.73491	18
43	.67837	.92331	1.0831	.73472	17
44	.67859	.92385	1.0824	.73452	16
45	.67880	.92439	1.0818	.73432	15
46	.67901	.92493	1.0812	.73413	14
47	.67923	.92547	1.0805	.73393	13
48	.67944	.92601	1.0799	.73373	12
49	.67965	.92655	1.0793	.73353	11
50	.67987	.92709	1.0786	.73333	10
51	.68008	.92763	1.0780	.73314	9
52	.68029	.92817	1.0774	.73294	8
53	.68051	.92872	1.0768	.73274	7
54	.68072	.92926	1.0761	.73254	6
55	.68093	.92980	1.0755	.73234	5
56	.68115	.93034	1.0749	.73215	4
57	.68136	.93088	1.0742	.73195	3
58	.68157	.93143	1.0736	.73175	2
59	.68179	.93197	1.0730	.73155	1
60	.68200	.93252	1.0724	.73135	0
′	Cos	Cot	Tan	Sin	′

132° (312°)   **(227°) 47°**

43° (223°)   **(316°) 136°**

′	Sin	Tan	Cot	Cos	′
0	.68200	.93252	1.0724	.73135	60
1	.68221	.93306	1.0717	.73116	59
2	.68242	.93360	1.0711	.73096	58
3	.68264	.93415	1.0705	.73076	57
4	.68285	.93469	1.0699	.73056	56
5	.68306	.93524	1.0692	.73036	55
6	.68327	.93578	1.0686	.73016	54
7	.68349	.93633	1.0680	.72996	53
8	.68370	.93688	1.0674	.72976	52
9	.68391	.93742	1.0668	.72957	51
10	.68412	.93797	1.0661	.72937	50
11	.68434	.93852	1.0655	.72917	49
12	.68455	.93906	1.0649	.72897	48
13	.68476	.93961	1.0643	.72877	47
14	.68497	.94016	1.0637	.72857	46
15	.68518	.94071	1.0630	.72837	45
16	.68539	.94125	1.0624	.72817	44
17	.68561	.94180	1.0618	.72797	43
18	.68582	.94235	1.0612	.72777	42
19	.68603	.94290	1.0606	.72757	41
20	.68624	.94345	1.0599	.72737	40
21	.68645	.94400	1.0593	.72717	39
22	.68666	.94455	1.0587	.72697	38
23	.68688	.94510	1.0581	.72677	37
24	.68709	.94565	1.0575	.72657	36
25	.68730	.94620	1.0569	.72637	35
26	.68751	.94676	1.0562	.72617	34
27	.68772	.94731	1.0556	.72597	33
28	.68793	.94786	1.0550	.72577	32
29	.68814	.94841	1.0544	.72557	31
30	.68835	.94896	1.0538	.72537	30
31	.68857	.94952	1.0532	.72517	29
32	.68878	.95007	1.0526	.72497	28
33	.68899	.95062	1.0519	.72477	27
34	.68920	.95118	1.0513	.72457	26
35	.68941	.95173	1.0507	.72437	25
36	.68962	.95229	1.0501	.72417	24
37	.68983	.95284	1.0495	.72397	23
38	.69004	.95340	1.0489	.72377	22
39	.69025	.95395	1.0483	.72357	21
40	.69046	.95451	1.0477	.72337	20
41	.69067	.95506	1.0470	.72317	19
42	.69088	.95562	1.0464	.72297	18
43	.69109	.95618	1.0458	.72277	17
44	.69130	.95673	1.0452	.72257	16
45	.69151	.95729	1.0446	.72236	15
46	.69172	.95785	1.0440	.72216	14
47	.69193	.95841	1.0434	.72196	13
48	.69214	.95897	1.0428	.72176	12
49	.69235	.95952	1.0422	.72156	11
50	.69256	.96008	1.0416	.72136	10
51	.69277	.96064	1.0410	.72116	9
52	.69298	.96120	1.0404	.72095	8
53	.69319	.96176	1.0398	.72075	7
54	.69340	.96232	1.0392	.72055	6
55	.69361	.96288	1.0385	.72035	5
56	.69382	.96344	1.0379	.72015	4
57	.69403	.96400	1.0373	.71995	3
58	.69424	.96457	1.0367	.71974	2
59	.69445	.96513	1.0361	.71954	1
60	.69466	.96569	1.0355	.71934	0
′	Cos	Cot	Tan	Sin	′

133° (313°)   **(226°) 4**